OLIVER EVANS,

THE WATT OF AMERICA.

A CATECHISM

OF

HIGH PRESSURE

OR

Non-Condensing Steam Engines.

INCLUDING THE

MODELING, CONSTRUCTING, RUNNING AND MANAGEMENT

OF

STEAM ENGINES AND STEAM BOILERS.

With Illustrations.

By STEPHEN ROPER,
ENGINEER.

PHILADELPHIA:
CLAXTON, REMSEN & HAFFELFINGER,
624, 626 & 628 Market Street.
1874.

Entered, according to Act of Congress, in the year 1873, by
STEPHEN ROPER,
in the Office of the Librarian of Congress, at Washington.

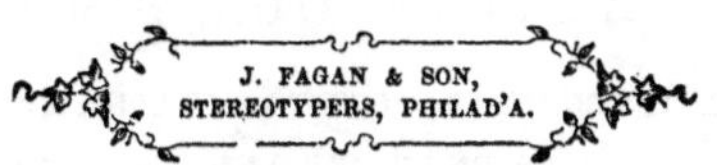

Selheimer & Moore, Printers,
501 Chestnut Street.

THIS VOLUME

IS

TO

ARTHUR ORR, D. M. P.,

AS A TRIBUTE OF RESPECT TO HIS EMINENT ABILITIES
AS AN ENGINEER AND MECHANIC.

INTRODUCTION.

THE writer of the following pages has made it his study to produce a work for the use of practical engineers, by furnishing them with a simple and clear explanation of each subject treated. In its preparation a great effort has been made to secure accuracy, and to render every subject plain and easy of comprehension, by treating it in a clear style, taking up each subject in regular order and examining it by the light of daily experience. The range of subjects comprehends everything directly connected with the steam-engine and steam-boiler. At the end of many of the articles, tables have been appended and examples introduced, to make explicit and distinct the principles set forth.

Many of the books heretofore published have been written by men of mere theory, for the purpose of instructing practical engineers in matters which the authors have never learned. Abstract theory can never meet the wants of practical men, nor instruct them in the discharge of their duties. But the writer does not wish to be considered an unbeliever in theories; because theories can never be dispensed with by any class of men, however unlearned, as theories aid us in the discovery of new facts and new truths. A book, however, to be useful to all classes of engineers, must give strong evidence of the author's experience as a practical engineer. The writer has had an experience of over thirty years with every description of engines and boilers, and in the preparation of this little book he has drawn almost entirely upon his own experience and observation. His aim has been to convey his meaning, to those for whom it is intended, by means of plain language, with familiar and practical illustrations, and to instruct those who are intrusted with the care and management of steam-engines and steam-boilers, so that they may understand with certainty whether they are deriving the greatest amount of practical advantage from the power they have generated. Hints and examples have also been given, intended to show how a great many practical improvements can be made by engineers and owners of steam-engines.

Another feature of the work, and one which it is believed will greatly benefit engineers of limited education, is that the use of decimals has been dispensed with, and fractions also, except those expressed by the common signs, ¼, ½, ¾, &c.

In order to render the book useful to those who employ or are employed about steam-engines, the writer has endeavored to embody all the necessary information relative to improvements in the construction, use and management of steam-engines and steam-boilers that have come into use up to the present time.

S. R.

CONTENTS.

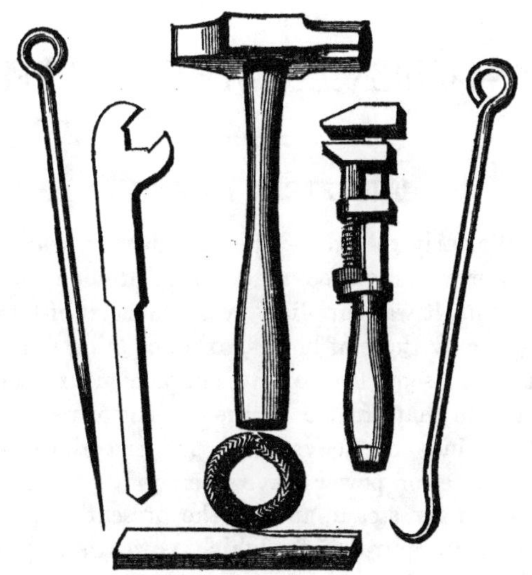

THE ENGINEER'S CHART.

A CATECHISM
OF
HIGH PRESSURE,
OR
NON-CONDENSING STEAM-ENGINES.

THE STEAM-ENGINE.

IN the early records of the human race, whenever power was required in order to provide for human wants, it was supplied by the muscles of human beings, or by those of horses, oxen, or other animals.

But men soon learned to supplement muscular power with that created by the descent of water, and by the winds of heaven. In modern times even these sources of power have been to a great extent superseded by steam, and at the present day there is hardly any purpose for which power is required, that is not furnished by the steam-engine.

Of all the efforts of human ingenuity known, perhaps none has monopolized so large a share of inventive genius as the steam-engine. No other object in the entire range of human devices has so irresistibly arrogated to itself the devotion of scientific men, as

2

the production of an artificial movement from the vapor of boiling water.

The progressive history of the invention commences with the days of Hero of Alexandria, two thousand years ago, and advances down through the labors of the Marquis of Worcester, of Savery and Papin, to the days of Watt, whose splendid improvements of the working principle of the steam-engine are well known.

In late years the improvement of the steam-engine has been steady, rather than remarkable in any one particular. It has advanced by improvement in construction, rather than by the development of any new principle. The facilities for the manufacture of steam-engines, and the amount of capital invested in that branch of mechanics, has increased very rapidly within the last few years.

Thousands of skilled workmen are continually inventing ingenious devices to replace the different parts of the engine now in use, and modifications are constantly resorted to to secure good construction and general improvement.

As might be expected, we find an infinite variety of constructions, some of inferior design, inferior workmanship, and inferior finish, whilst others have splendid symmetrical proportions, elaborate finish, and beautiful and simple mechanism.

But still there appears to be a wide field open for experiment and improvement in the steam-engine, since it is asserted by persons competent to judge, that first-class engines yield less than ten per cent. of the work stored up in good fuel, and the average engines probably utilize less than four per cent. of the fuel.

The best steam-engine, apart from its boiler, has about five-sixths of the efficiency of a perfect engine. The remaining sixth is lost through waste of heat by radiation and conduction externally, and by condensation and friction externally. It is to improvements in these points that inventors must turn their attention.

WATER.

Q. What are the constituent parts of fresh-water?

A. Oxygen and hydrogen.

Q. In what proportion?

A. Oxygen, 89 parts by weight, and by measure 1 part; hydrogen, by weight 11 parts, and by measure 2 parts.

Q. How many cubic feet of fresh-water make 1 ton?

A. 36 cubic feet.

Q. How many cubic feet of sea-water make 1 ton?

A. 35 cubic feet.

Q. How do you account for the difference in weight between sea- and fresh-water?

A. Sea-water is more dense.

Q. How much does a cubic foot of water weigh?

A. 62½ pounds.

Q. How many gallons in a cubic foot of water?

A. 6¼ gallons.

Q. How much does a cubic inch of water weigh?

A. About ½ an ounce.

Q. How much does a cylindrical foot of water weigh?

A. 49 pounds.

Q. What is the weight of a cylindrical inch of water?

A. Nearly ½ an ounce.

Q. How much does a cubic foot of ice weigh?

A. 68½ pounds.

Q. What is the boiling-point for water?

A. 212° Fah.

Q. What is the freezing-point of water?

A. 32° Fahrenheit.

Q. What is the difference between water, ice, and steam?

A. Water is a fluid, ice is a solid, steam is a vapor.

Q. What is the difference in volume between water and steam at the pressure of the atmosphere?

A. 1700; that is to say, that any given quantity of water will make 1700 times that amount of steam at a pressure of 15 pounds to the square inch, or atmospheric pressure.

Q. At what temperature does water attain its greatest density?

A. 39° Fah.

Q. Is water compressible?

A. No: it presses in every direction and finds its level.

Q. What is the centre of pressure of a column of water?

A. $\frac{2}{3}$ of its depth from the surface.

Q. At what degree of temperature does the water boil in a vacuum?

A. 98° Fah.

Q. At what degree of temperature does water become solid?

A. 32° in the open air.

Q. Does water expand in freezing?

A. Yes.

Q. Do all fluids expand with heat?

A. Yes; all fluids expand with heat, and contract with cold down to 40° Fah.

Q. Are all waters equal for the production of steam?

A. No; sea-water, or other waters holding salt in solution, require a higher temperature to produce steam of the same elastic force.

Q. How is the gravity of water ascertained?

A. By means of a hydrometer.

Q. If water be boiled in an open vessel, can the temperature of the water be raised above 212° Fah.?

A. No; as all the surplus heat which may be applied passes off with the steam.

Q. How do you explain the theory of ebullition, or boiling water?

A. In this way; that in metals heat is communicated by the conducting properties they possess, but in liquids heat is communicated by a separation of particles.

Q. If heat be applied to the top of a vessel containing water, will ebullition take place?

A. No; as very little heat will be communicated to other parts of the vessel, and the water will never boil.

Q. Is it a common thing in steam-boilers to indicate a larger supply of water than that which really exists in the boiler?

A. Yes; as the steam is forming in the water it rises in bubbles to the surface, which by their bulk displace a great amount of water, indicating a great rise at the gauge-cocks.

Q. What is the vaporization of water?

A. It is the conversion of water, as a liquid, into vapor as steam.

Q. What is solidification?

A. It is a change which water undergoes from a liquid to ice as a solid.

Q. Does the density of water increase with its depth?

A. Yes; there is a theoretical depth at which water would become as dense as iron.

TABLE

Showing the Weight of Water in Pipes of Various Diameters One Foot in Length.

Diameter in Inches.	Weight in Pounds.	Diameter in Inches.	Weight in Pounds.	Diameter in Inches.	Weight in Pounds.
3	3	12¼	51	23	180¼
3¼	3½	12½	53¼	23½	188¼
3½	4¼	12¾	55½	24	196¼
3¾	4¾	13	57½	24½	204½
4	5½	13¼	59¾	25	213
4¼	6¼	13½	62¼	25½	221½
4½	7	13¾	64½	26	230½
4¾	7¾	14	66¾	26½	239½
5	8½	14¼	69¼	27	248½
5¼	9¼	14½	71½	27½	257¾
5½	10¼	14¾	74¼	28	267¼
5¾	11¼	15	76¾	28½	276¾
6	12¼	15¼	79¼	29	286½
6¼	13¼	15½	82	29½	296½
6½	14½	15¾	84½	30	306¾
6¾	15½	16	87¼	30½	317¼
7	16¾	16¼	90	31	327½
7¼	18	16½	92¼	31½	338¼
7½	19¼	16¾	95½	32	349
7¾	20½	17	98½	32½	360
8	21¾	17¼	101½	33	371¼
8¼	23¼	17½	104½	33½	382½
8½	24½	17¾	107½	34	394
8¾	26	18	110½	34½	405¾
9	27½	18¼	113½	35	417½
9¼	29¼	18½	116½	35½	429½
9½	30¾	18¾	119¾	36	441¾
9¾	32½	19	123	36½	454
10	34	19¼	126¼	37	466½
10¼	35½	19½	129½	37½	479¼
10½	37½	19¾	132	38	492¼
10¾	39¼	20	136¼	38½	505¼
11	41¼	20½	143¼	39	518½
11¼	43¼	21	150¼	39½	531¾
11½	45	21½	157½	40	545½
11¾	47	22	165		
12	49	22½	172½		

AIR.

Q. What are the constituent parts of air, or what does air consist of?

A. It consists, by volume, of oxygen 21 parts, of nitrogen 79 parts, and by weight, of oxygen 77 parts, and of nitrogen 23 parts.

Q. Does any other gas enter into the constituent parts of air?

A. Yes; in 1,000,000 parts of air there are 4 parts of carbonic acid.

Q. How many cubic feet in 1 pound of air?

A. 13,817 cubic feet.

Q. What is the weight of 1 cubic foot of air at the surface of the earth, temperature 34° Fah.?

A. 527 grains, or ¼ ounce avoirdupois.

Q. What is the difference in weight between air and water?

A. Air is 829 times lighter than water.

Q. What is the mean weight of a column of air 1 inch square and 45 miles high?

A. 15 pounds.

Q. Is the pressure of the air the same at all altitudes?

A. No; at 7 miles above the surface of the earth the air is 4 times lighter than at the earth's surface; at 14 miles, 16 times; at 21 miles, 64 times.

Q. How much air does it require to consume 1 pound of coal?

A. 18 pounds, or 240 cubic feet.

Q. Is it possible to construct an air-engine of any great power?

A. No; as air, like all other gases, expands but one volume for each 493° of temperature through which it is raised; and in order to double its volume, we must raise it 493° more, which will bring it to a temperature of 986° Fah., which is entirely too high for practical purposes. Even if air could be worked at this temperature, it would be necessary to have a feed-pump of nearly the capacity of the engine, which would be very cumbrous, and the engine itself would have to be more than twice the size of a steam-engine of the same power.

Q. Is it possible to obtain much power from air at a lower temperature?

A. Yes; but as we lower the temperature, we must increase the capacity of the engine and pump, which will become very bulky and expensive.

Q. Is the expansion of air, like all other elastic fluids, uniform at all temperatures?

A. Yes; the following table will show the rate of expansion of and according to temperature:

TABLE.

EXPANSION OF AIR BY HEAT. SHOWING THE INCREASE OF BULK IN PROPORTION TO THE INCREASE OF TEMPERATURE.

Fahrenheit.		Bulk.	Fahrenheit.		Bulk.
Temp. 32	Freezing-point.	1000	Temp. 75		1099
" 33		1002	" 76	Summer heat...	1101
" 34		1004	" 77		1104
" 35		1007	" 78		1106
" 36		1009	" 79		1108
" 37		1012	" 80		1110
" 38		1015	" 81		1112
" 39		1018	" 82		1114
" 40		1021	" 83		1116
" 41		1023	" 84		1118
" 42		1025	" 85		1121
" 43		1027	" 86		1123
" 44		1030	" 87		1125
" 45		1032	" 88		1128
" 46		1034	" 89		1130
" 47		1036	" 90		1132
" 48		1038	" 91		1134
" 49		1040	" 92		1136
" 50		1043	" 93		1138
" 51		1045	" 94		1140
" 52		1047	" 95		1142
" 53		1050	" 96	Blood heat......	1144
" 54		1052	" 97		1146
" 55		1055	" 98		1148
" 56	Temperate.......	1057	" 99		1150
" 57		1059	" 100		1152
" 58		1062	" 110	Fever heat 112	1173
" 59		1064	" 120		1194
" 60		1066	" 130		1215
" 61		1069	" 140		1235
" 62		1071	" 150		1255
" 63		1073	" 160		1275
" 64		1075	" 170	Spirits boil 176	1295
" 65		1077	" 180		1315
" 66		1080	" 190		1334
" 67		1082	" 200		1364
" 68		1084	" 210		1372
" 69		1087	" 212	Water boils.....	1375
" 70		1089	" 302		1558
" 71		1091	" 392		1739
" 72		1093	" 482		1919
" 73		1095	" 572		2098
" 74		1097	" 680		2312

HEAT.

Q. What is heat?

A. It is a species of motion, or one form of mechanical power.

Q. Is heat capable of producing power?

A. Yes; and power is capable of producing heat.

Q. With any given degrees of temperature, and any given expenditure of heat, will the amount of power generated be the same?

A. Yes.

Q. What is specific heat?

A. Specific heat of a substance is an expression for the quantity of heat in any given weight of it at certain temperatures.

Q. What is sensible heat?

A. That which is sensible to the touch.

Q. What is latent heat?

A. It is that which a body absorbs in changing from a solid to a fluid state.

Q. What is meant by the radiation of heat?

A. It is the effect produced by the direct rays of a hot body through space.

Q. Is the latent heat in steam uniform at all temperatures?

A. No; neither is the total amount of heat the same at all temperatures.

Q. Does the total and latent heat in steam increase with the pressure?

A. No; as the sensible heat increases the latent heat diminishes.

Q. Can the absolute heat of any body be determined?

A. Only by the relative heat of other bodies at the same temperature.

Q. What is combustion?

A. It is an energetic combination of gases, or a mutual neutralization of opposing electricities.

Q. Does the quantity of heat in any body vary with the temperature?

A. Yes; the temperature of steam rises with the pressure.

Q. What method is adopted to determine temperatures so high that no thermometer can give a reliable result?

A. We take a body, such as platinum, and place a mass of this metal in a blast furnace, and when the mass has acquired the temperature of the furnace we transfer it to a vessel containing a known weight of water. We can then observe the rise of temperature by means of an ordinary thermometer.

Q. How do you determine the specific heat of different bodies?

A. By mixing different substances together at different temperatures, and noting the temperature of the mixture. For instance, when water is at 100° and mercury at 40°, the mixture will be at 80°; the 20° lost by the water causes a rise of 40° in the mercury.

Q. What is a unit of heat?

A. The unit of heat is the amount of heat required to raise the temperature of 1 pound of water 1°, or from 32° to 33° Fah.

Q. What is the mechanical equivalent of heat?

A. The power necessary to raise 1 pound 772 feet high.

Q. What is the capacity of any body for heat?

A. It is the relative power of a body in receiving and retaining heat at any given temperature.

Q. What is the reflection of heat?

A. It is the passage of heat from one surface to another, or into space.

Q. What is the communication of heat?

A. It is the passage of heat to different bodies with different degrees of velocity.

Q. What is the transmission of heat?

A. It is the passage of heat through different degrees of intensity.

Q. How much heat does a pound of water receive in passing from a liquid at 212° Fah. to a vapor at 212°?

A. It receives as much heat as would raise it 966° if the heat was sensible as well as latent.

Q. What is conversion of heat?

A. It is the transfer or diffusion of heat in a fluid mass by means of its particles.

Q. Will water boil in a vacuum with less heat than under the pressure of the atmosphere?

A. Yes; in a vacuum water boils at 98° to 100°, according as the vacuum is perfect.

Q. Does water give out heat in freezing?

A. Yes; water in freezing gives 140° of heat.

Q. At what degree of temperature do fluids evaporate in a vacuum?

A. From 100° to 120° below the boiling-point.

Q. What is a thermal unit?

A. It is the quantity of heat required to raise 1 pound of water 1°, the water being at its maximum density (= 39° Fah.).

Q. How can the amount of heat stored up in any fluid, that might be utilized by perfect mechanism, be shown?

A. It must be represented by a fraction, the numerator of which is the temperature of the fluid while doing its work, and the denominator the temperature of the fluid when entering a vessel, to be measured from absolute zero.

Q. Does the nature of the medium upon which heat acts in the production of power make any difference?

A. No; whether water, air, metal, or any other substance is the medium, it is immaterial, except so far as the agent is convenient and manageable; and just in the proportion in which power is generated, so will heat disappear.

TABLE.

The following Table will show the Temperature required to Ignite different Substances.

Name of Substance.	Temp. of Ignition.
Phosphorus	140° Fah.
Bisulphide of Carbon	300° "
Fulminating Powder	374° "
Fulminate of Mercury	392° "
Gun-Cotton	428° "
Nitro-Glycerine	494° "
Rifle-Powder	550° "
Forced Gunpowder	563° "
Picrate Powder for Torpedoes	570° "
Charcoal from Willow Wood	660° "
Picrate Powder for Cannon	715° "
Very Dry Pine Wood	800° "
Dry Oak Wood	900° "
Steam, 100 lbs. per square inch pressure	332° "
Steam, 200 " " " " "	387° "
Steam, 240 " " " " "	403° "

THE THERMOMETER.

Q. What is an absolute zero?

A. It is the point at which heat-motion is supposed to cease entirely, estimated to be 461° Fah. below the zero of the common scale.

Q. How is the zero of the common scale fixed?

A. It is fixed by mixing salt with snow until the mercury in the tube falls 32° below the freezing-point for water.

Q. What do you mean by Fahrenheit?

A. I mean Fahrenheit's scale or thermometer, the one generally used in this country.

Q. What do you mean by Centigrade?

A. I mean the Centigrade scale or thermometer, by which they measure temperatures in France.

Q. What do you mean by Reaumur?

A. I mean Reaumur's rule or thermometer that is commonly used in Germany.

Q. What is the difference between Fahrenheit's, Centigrade, and Reaumur's rules?

A. Fahrenheit's zero is 32° below freezing, boiling-point of water 212°; Centigrade, zero 32° or freezing, boiling-point 100°; Reaumur's, zero 32° or freezing, boiling-point 80°. Hence Fahrenheit 180°, Centigrade 100°, Reaumur 80°.

Q. What are fixed temperatures?

A. One the melting-point of ice, and the other the boiling-point of pure water.

Q. Why do you call these fixed temperatures?

A. Because it is impossible to raise the temperature of ice above 32° Fah., and no amount of heat will raise boiling water above a temperature of 212° Fah., if contained in an open vessel.

Q. Does the thermometer indicate the amount of heat in any body?

A. No; only the changes in temperature.

Q. What is a differential thermometer?

A. An instrument that measures minute differences of temperature.

COMPARATIVE SCALE OF ENGLISH, FRENCH AND GERMAN THERMOMETERS.

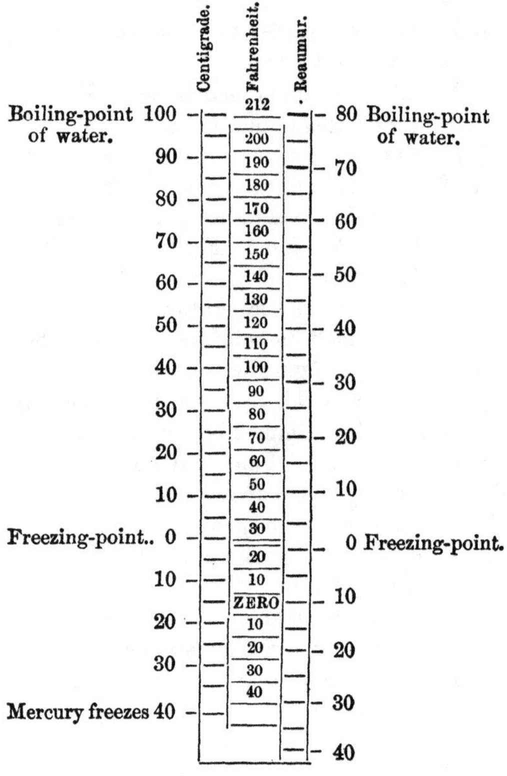

STEAM.

Q. What is steam?

A. Steam is vapor arising from water at a temperature of 212° Fah.

Q. What are the most prominent properties possessed by steam?

A. First, its high expansive force; second, its property of condensation; third, its concealed or undeveloped heat.

Q. By whom were the expansive properties of steam discovered?

A. By Hornblower, who obtained a patent for his invention in 1781.

Q. From what cause does the expansive force of steam arise?

A. From the absence of cohesion between the particles of water.

Q. How is the condensation of steam effected?

A. By the abstraction of its temperature.

Q. What is the difference in volume between water and steam at a temperature of 212° Fah.?

A. 1700; that is to say, any given quantity of water converted into steam at the pressure of the atmosphere or 212° Fah., will present a volume 1700 times greater than its original bulk.

Q. Can steam mix with air?

A. Not while its pressure exceeds that of the atmosphere.

Q. If a cylinder be filled with steam at a pressure of 15 pounds to the square inch, will the air be expelled from the cylinder?

A. Yes.

Q. Now as the existence of the steam depends upon its temperature, by abstracting that temperature will the steam assume the state due to its reduced temperature?

A. Yes; by immersing the cylinder in cold water the steam contained therein is condensed into water.

Q. Now, as the water cannot occupy the volume which it did under its former temperature, what is the result?

A. As the steam in the cylinder is condensed by the abstraction of its heat, the result will be a vacuum.

Q. What is the amount of latent or concealed heat that exists in steam?

A. The latent heat of steam, though showing no effect on the thermometer, may be easily shown by placing 5½ pounds of water in a vessel at 32° Fah., and admitting steam through a pipe from another vessel until the temperature of the water in the first vessel is raised to the boiling-point. It will then be discovered that the weight of the water in the vessel is 6½ pounds.

Q. How do you account for this additional pound of water?

A. It is the result of the admission of one pound

of steam to the vessel; and this pound of steam, while retaining its own temperature of 212°, has raised 5½ pounds of water 180°, or an equivalent to 990°, and including its own temperature we have 1202°, which the pound of steam must have possessed at first.

Q. Is the sum of latent and sensible heat of steam nearly constant?

A. Yes; and does not vary much from 1200°.

Q. If a known volume of steam of a certain pressure be made to occupy ½ that volume, is its elastic force doubled?

A. Yes; the same pressure is exerted within ½ of its original capacity.

Q. What do you mean by steam pressure?

A. I mean the initial elastic force of the steam, which is always the same in equal weights of steam.

Q. Does the elasticity of steam increase with an increase of temperature?

A. Yes, but not in the same ratio; because if steam is generated from water at a temperature which gives it the pressure of the atmosphere, an additional temperature of 38° will give it a temperature of two atmospheres, and a still further addition of 42° will give it a pressure of 4 atmospheres.

Q. Why is it that every additional degree of temperature between 40 and 50 doubles the pressure?

A. The heat in generating the steam has to overcome the attraction among the particles of water, and likewise of the gravity of the water itself; and as the water becomes rarefied by heat, the attraction and gravity become diminished, and an additional temperature does not have to contend with the same resistance as the one that preceded it.

Q. Is steam in itself invisible?

A. Yes; and it only becomes visible by loss of temperature, as when a jet is discharged into the open air, and is then seen in the form of vapor.

Q. In treating of steam, why are the terms heat and caloric used?

A. The term heat is understood as expressing sensible heat, while the term caloric expresses every conceivable existence of temperature.

Q. Is steam produced from impure water equal in density to steam from pure water?

A. No; steam from pure water, at a temperature of 212° Fah., supports a column of mercury of 30 inches, while steam from sea or impure water, at the same temperature, supports only 22 inches.

Q. Can the heat of steam be raised to a very high temperature?

A. Yes; steam can be heated to nearly a red heat, but not while it is held in contact with water.

Q. Does the temperature of steam at ordinary pressure contain heat enough to ignite wood?

A. Not without the intervention of some other substance, such as linseed oil, greasy rags, or iron turnings.

Q. How do you explain that?

A. Because we know that the temperature of superheated steam is only about 400° Fah., and it requires more than double that intensity of heat to ignite wood. *See Table under the head of Heat.*

Q. What is liquefaction?

A. It is the condensing of steam through the abstraction of heat.

Q. Do you know of any reliable rule for connecting the temperature and elastic force of saturated steam?

A. No; various formulas have been at different times introduced for the purpose of deducing the elastic force of saturated steam, but without any very reliable or satisfactory results.

Q. Can you tell the number of superficial feet of steam-pipe necessary to warm any room or number of rooms?

A. One superficial foot of steam-pipe to 6 superficial feet of glass in the windows, or 1 superficial foot of steam-pipe for every 100 square feet of wall, roof or ceiling, or 1 square foot of steam-pipe to 80 cubic feet of space; 1 cubic foot of boiler is required for every 1500 cubic feet of space to be warmed. One horse-power boiler is sufficient for 40,000 cubic feet of space.

TABLE.

Showing the Temperature of Steam at different Pressures, from 15 pounds per Square Inch to 100 pounds, and the quantity of Steam produced from a Cubic Inch of Water according to Pressure.

Total Pressure of Steam in lbs. perSquare Inch.	Corresponding Temperature of Steam to Pressure.	Cubic Inches of Steam from a Cubic Inch of Water according to Pressure.	Total Pressure of Steam in lbs. perSquare Inch.	Corresponding Temperature of Steam to Pressure.	Cubic Inches of Steam from a Cubic Inch of Water according to Pressure.
15	212.8	1669	58	292.9	484
16	216.3	1573	59	294.2	477
17	219.6	1488	60	295.6	470
18	222.7	1411	61	296.9	463
19	225.6	1343	62	298.1	456
20	228.5	1281	63	299.2	449
21	231.2	1225	64	300.3	443
22	233.8	1174	65	301.3	437
23	236.3	1127	66	302.4	431
24	238.7	1084	67	303.4	425
25	241.0	1044	68	304.4	419
26	243.3	1007	69	305.4	414
27	245.5	973	70	306.4	408
28	247.6	941	71	307.4	403
29	249.6	911	72	308.4	398
30	251.6	883	73	309.3	393
31	253.6	857	74	310.3	388
32	255.5	833	75	311.2	383
33	257.3	810	76	312.2	379
34	259.1	788	77	313.1	374
35	260.9	767	78	314.0	370
36	262.6	748	79	314.9	366
37	264.3	729	80	315.8	362
38	265.9	712	81	316.7	358
39	267.5	695	82	317.6	354
40	269.1	679	83	318.4	450
41	270.6	664	84	319.3	346
42	272.1	649	85	320.1	342
43	273.6	635	86	321.0	339
44	275.0	622	87	321.8	335
45	276.4	610	88	322.6	332
46	277.8	598	89	323.5	328
47	279.2	586	90	324.3	325
48	280.5	575	91	325.1	322
49	281.9	564	92	325.9	319
50	283.2	554	93	326.7	316
51	284.4	544	94	327.5	313
52	285.7	534	95	328.2	310
53	286.9	525	96	329.0	307
54	288.1	516	97	329.8	304
55	289.3	508	98	330.5	301
56	290.5	500	99	331.3	298
57	291.7	492	100	332.0	295

THE ENGINEER.

THE duties of an engineer intrusted with the care and management of a steam-engine and steam-boilers, are of more importance than would appear at first sight, as the mechanics of any other class might be ignorant or neglectful of duty without subjecting themselves or their employers to any danger or much inconvenience; but ignorance or neglect of duty on the part of an engineer, might at any time result in great destruction to life and property.

Steam-boilers require constant care and attention, and when in charge of careful and intelligent engineers are perfectly safe, when well-constructed; but when their management falls into the hands of reckless and ignorant men, they become a source of constant danger to their owners and the public. For the above reasons any engineer who may offer to take charge of an engine and boiler ought to be able to show that he fully understands the duties and responsibilities involved in their care and management. He ought to be able to calculate safety-valve lever examples, and thoroughly understand the principles involved. He should also be able to calculate the pressure required to burst a boiler when all the dimensions are given. He should understand the difference between longitudinal and curvilinear strains, the difference in value between single- and double-riveted seams, and the comparative strength of the shell, flues, and other parts of the boiler. He should also fully understand the causes which tend to produce explosions, and be conversant with all the details of construction, use and management of steam-boilers. But, above all, he should be a man of good sound sense and intelligence.

Now, it requires time, study and experience to acquire the above fund of information; and if the engineer should qualify himself accordingly, will his intelligence be appreciated and his services be remunerated by steam users? Yes; undoubtedly they will, as there seems to be a steady and increasing demand for skilled workmen in this country in every branch of mechanics — and in none more so than in steam-engineering. Perhaps there may be some steam users who do not desire the services of a good engineer, for the reason that they think he might interpose an intelligent remonstrance against their recklessness in the use and management of their steam-boilers, or he might decline to gratify their avarice by allowing himself to be degraded to the position of man-of-all-work. Not being familiar with the properties of steam themselves, many owners of steam-boilers seem to think that almost any man — no matter how ignorant he may be — can in a short time qualify himself for the position of an engineer; it not seeming to occur to them that there is no place where experience and intelligence are of so much importance as they are in the care and management of steam-boilers.

It is also a frequent assertion of some steam users that a great many engineers are incapable. They seem to forget that the same thing might be said of the members of any other trade or profession. But it might be said, without fear of contradiction, that engineers in general are as capable and intelligent as any other class of mechanics. Even if the above assertion is true — that a great many engineers are incapable — it might be pertinent to the question to ask: Are they wholly to blame? or does some of the responsibility rest with the owners of steam-engines and steam-boilers, who have been for years encouraging ignorance and incapacity by employing incompetent men, from motives of false economy?

Under such circumstances, it is not to be wondered at that a great many errors have found their way into the business; but the remedy for nearly all of them is within reach of the engineers themselves — more particularly the young men. They must educate themselves, so that their opinion on all questions connected with steam-engineering will give strong evidence that they fully understand the principles and practice of their profession, then their opinions will command respect, and their services be correspondingly remunerated. They must remember that it is not the trade that elevates the man, but it is rather the man that dignifies the pursuit or calling;

and that muscular power, though very good in its place, is not the most essential requisite of an engineer, but that the cultivation of the mind is the first step towards eminence in any trade or profession.

It is true that in steam-engineering, as in all other trades, there are a great many men who are totally unfit for the business — men that, perhaps, would succeed, to a certain extent, in some other pursuit, but who become a failure, and often a reproach to the profession they have adopted, simply for the reason that they have made a mistake in the selection of a suitable trade.

Although no modern writer on steam or the steam-engine has made any allusion to engineers, or called public attention to the importance of their education to steam users and the public, still it cannot be denied that engineers have made very creditable progress in the acquirement of knowledge connected with their business within the past ten years. But yet there seems to be ample room for improvement on the part of both engineers and their employers, and that such change will soon come there is no reason to doubt. For improvement in steam-engines and steam-boilers will also require more able management; and intelligent steam users will begin to see that the most important essentials to the economical working of the steam-boiler and steam-engine are care, skill and intelligence, and these qualifications will liberally repay all the money expended in securing them.

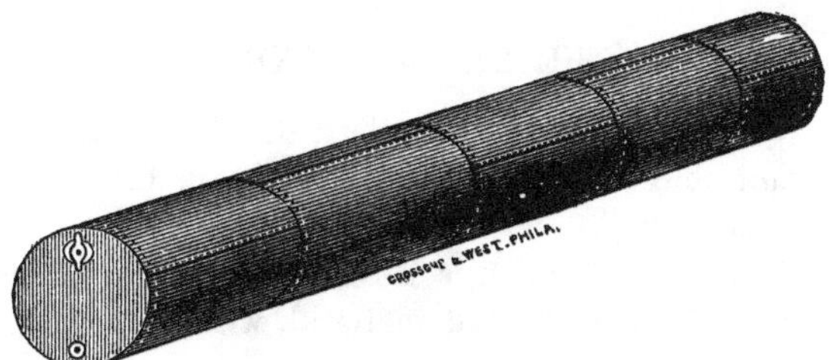

PLAIN CYLINDER BOILER.

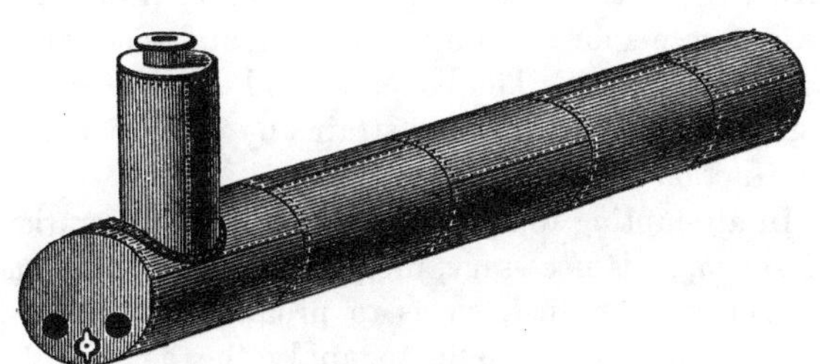

FLUE BOILER.

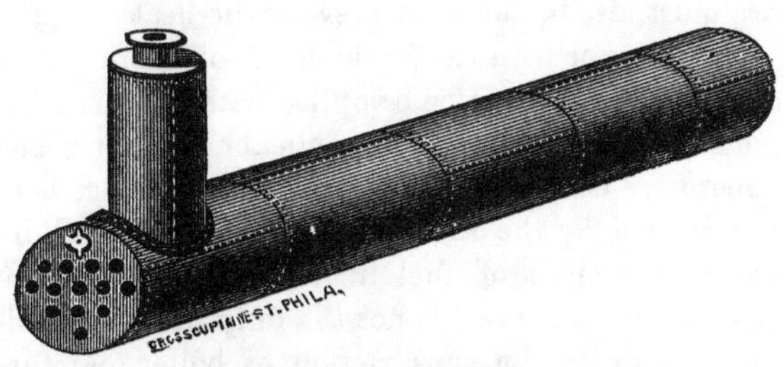

TUBULAR BOILER.

THE STEAM-BOILER.

In the choice of steam-boilers, the three most important objects to be attained are safety, durability, and economy.

To secure safety, it is necessary that the boiler should be made of good material, with good workmanship.

To secure durability, the boiler ought to be constructed so as to give the greatest facilities and easiest access for cleaning, repairing, and renewal of any of its parts. The boiler should also be so designed as to avoid unequal strains by expansion and contraction as far as possible.

In attempting to secure economy in the generation of steam, it is necessary, first, to secure perfect combustion of the fuel, so as to produce the greatest amount of heat; second, to apply the heat in the very best manner to the boiler, so as to heat the water in the most rapid manner possible; third, great care must also be taken to prevent the heat escaping by radiation, or with the products of combustion. If these three conditions be complied with, our arrangements will be of the most economical character. The evaporative efficiency of any boiler and furnace is to be measured by the amount of water evaporated by any given weight of fuel in a given time. Mere waste of fuel, however, is not the only defect attendant upon an inferior construction of boiler and furnace. Where these are not of the best kind, they must be of larger size in order to do the required amount of work; the grate-surface must be larger, and more air must be needlessly raised to a higher temperature, thus carrying off a large amount of heat in the waste products of combustion; all of which involves increased outlay of capital and larger running expenses.

Many of the defects of modern boilers might be attributed, first, to the fact that some of the inventors or designers seem to be partly, if not totally, ignorant of the first principles of mechanical science; second, to competition between boiler-makers themselves, in their efforts to undersell each other, consequently they have to deceive purchasers and steam users by magnifying small boilers into large ones. Therefore, when the boiler comes to be tested, its evaporative powers are found to be lacking, the fuel has to be burned under a sharp draught, and instead of the best results, the worst are obtained.

In regard to the metal of the boiler itself, it is a well known fact that the thicker the iron is, and the poorer its conducting qualities, the greater will be the amount of heat that will be lost or wasted; when by using a superior quality of iron, one whose tensile strength and conducting powers are both very great, we lessen the resistance to the passage of the heat from the furnace to the water and greatly increase the economy of the boiler. It is well known to engineers that some qualities of iron are two and a half times stronger than others; consequently, if we make a boiler of poorer iron than would be as strong as $\frac{1}{4}$ inch of the best iron, we should have to use plates $\frac{5}{8}$ of an inch thick. Even then a heavy boiler would be weaker than the light one, from the fact that the heavy plates would sustain great injury in the making. In point of economy and durability, the light boiler would be far superior to the heavy one.

Every attempt to lessen the first cost of a boiler by diminishing the heating- and grate-surface is, to a certain extent, carrying out the principle of "penny wise and pound foolish."

An engine extra large for the work to be done causes a loss of fuel, whilst a boiler extra large for the work to be done makes a great saving.

A boiler taxed to its full evaporative capacity will evaporate from 5 to 6 pounds of water to 1 pound of coal. Double the size of the boiler, and you will get the same amount of steam with 35 to 40 per cent. less fuel.

The quantity of water evaporated depends not only on the heating-surface, grate-surface, and draught, but also on the quantity of air which passes through the furnace in a given time.

For instance: a locomotive boiler burning 10 pounds of coal on each square foot of grate-surface in an hour, will evaporate about 8 pounds of water for each pound of coal. The same boiler, running at a high speed, and burning 75 pounds of coal on each square foot of grate-surface, will evaporate 7 pounds of water for each pound of coal burned. Here is a vast difference in the total amount of evaporation,—each pound of coal produces less steam in the proportion of 9 to 7 pounds.

So that it will be seen that the economy of fuel in one case is very evident; in the other, it will be seen that the waste of fuel under a forced draught is very great.

A boiler may generate steam with great economy, but, owing to the steam being wasted by improper application to the engine, the result is unsatisfactory, and the boiler unjustly blamed. A boiler that carries out water with its steam may show a large evaporation, but the steam being wet, is almost useless in the engine.

The maximum evaporative capacity of any boiler is the amount of water it would evaporate in 10 hours from a temperature of 60° Fah., with good fuel, good draught, and the steam allowed to escape as fast as generated.

Steam-boilers are of almost every kind and description: cylinder, tubular, and flue; upright, horizontal, inclined, and patent.

CYLINDER BOILERS.

Q. What advantage do cylinder boilers possess over other boilers?

A. First, the old plain cylinder boiler is light, is easy to clean and repair; second, it is less dangerous, and requires less attention than any other kind of boilers now in use.

Q. What are the disadvantages of the cylinder boiler, or why is it so fast passing out of use, more particularly in cities?

A. First, the cylinder boiler, on account of its extreme length, is unsuited to cities, where space is so limited and of great value; second, because it requires a greater amount of fuel to evaporate a certain amount of water than any other boiler now in use; third, because it takes so long to raise steam from cold water.

Q. What do you believe to be the most economical length for a cylinder boiler?

A. It is found by experience that the length of a cylinder boiler should never exceed 7 times its diameter.

Q. What might be considered a horse-power in a cylinder boiler?

A. The term horse-power, as applied to the steam-boiler, has no definite application. It has been customary to fix on some unit of heating- and grate-surface as the unit of horse-power for a boiler; and that unit is about 15 square feet of heating-surface, and $\frac{3}{4}$ of a square foot of grate-surface to the horse-power in cylinder boilers.

Q. What would you call a fair evaporation of water in a cylinder boiler?

A. 6 pounds of water to one pound of coal.

Q. How should cylinder boilers be set?

A. Cylinder boilers, or any other boiler, should have an incline of 1 inch to every 20 feet towards the end where the blow-off is situated. In case cylinder boilers are extremely long, they ought to be supported in the centre with a midfeather.

FLUE BOILERS.

Q. What are the advantages and disadvantages of flue boilers?

A. The advantage is that they occupy less room than the cylinder, from the fact that they present a greater amount of heating-surface. The disadvantages are, first, they are heavier than the cylinder; second, they cost more; third, they are difficult to repair or clean; fourth, they are very dangerous on account of the liability of the flues to collapse.

Q. In making flue boilers, what proportion should be carried out in regard to the flues?

A. They should always be made of small diameter, not more in any case than 12 or 14 inches, for the reason that small diameters give them greater strength.

Q. In case it should be necessary to make flues of large diameter, say 18 or 20 inches, what is the best plan of constructing them?

A. The flues should be made with *butt-joints,* and of heavier iron than the shell of the boiler.

Q. What is a butt-joint?

A. It is a joint formed by bringing the ends of two sections of the flue together and overlapping them on the outside and inside with separate pieces of iron. This plan of construction gives them additional strength, and makes them less liable to collapse.

Q. What is a collapse?

A. It is a crushing in of a flue or tube by external pressure.

Q. What would be considered a fair allowance of grate- and heating-surface to the horse-power in flue boilers?

A. From 14 to 15 square feet of heating-surface, and $\frac{3}{4}$ square foot of grate-surface.

Q. What might be called fair evaporation in a flue boiler?

A. 7 pounds of water to 1 pound of coal.

Q. What length do you consider most economical for flue boilers?

A. The length of flue boilers should be about 5 times their diameter.

TUBULAR BOILERS.

Q. What advantage does a tubular boiler possess over the cylinder and flue boilers?

A. The tubular takes up less room, generates steam more rapidly, and requires less fuel; and the tubes are less dangerous than flues on account of their small diameter and great strength.

Q. What are the disadvantages of tubular boilers?

A. First, the tubular boiler is heavy, the first cost of them is more than either the flue or the cylinder; second, they are difficult to repair, and almost impossible to clean; third, they require great care and attention to prevent the water from becoming low and exposing the tubes to the action of the fire.

Q. What would you consider the proper length for a tubular boiler?

A. The length of a tubular boiler should be about 4 times its diameter.

Q. What would you consider fair evaporation in a tubular boiler?

A. The evaporation of 9 pounds of water to 1 pound of coal.

Q. What might be called a fair allowance of heating- and grate-surface to the horse-power in a tubular boiler?

A. 14 square feet of heating-surface, ½ square foot of grate-surface. But in the tubular, as in all other boilers, the horse-power of the boiler depends upon the following conditions: first, the grate-surface; second, the heating-surface; third, draught; fourth, the qualities of the fuel used; fifth, the conducting powers of the iron; and sixth, the proper application of the steam after it leaves the boiler.

DOUBLE-DECK BOILERS.

Q. What do you mean by a double-deck boiler?

A. I mean a tubular and cylinder connected together by what are called water-necks — the cylinder being uppermost and used as a steam-dome.

Q. What are the advantages of the double-deck boiler?

A. A double-deck boiler occupies less space than any of the three former kinds of boilers; it steams very economically, and is also more safe than the single tubular, because the lower or tubular cylinder is entirely free from water. There is very little danger of the water becoming low in this kind of a boiler. The ends of the tubes are not exposed to the action of the fire, as in the single tubular, as the draught passes under the boiler and returns through the tubes, and re-returns between the tubular and cylinder. This boiler seems to have no objectionable features, with the exception that the lower or tubular section is impossible to clean. It might be said also that the boiler is heavy and expensive; but it presents an immense amount of heating-surface.

LOCOMOTIVE BOILERS.

Q. What are the advantages and disadvantages of the locomotive boiler?

A. The advantages are: the locomotive form of the boiler is compact and powerful, and, when well proportioned to its work, is very economical for factory use; it occupies small space and generates steam very rapidly. Its disadvantages are: first, it is expensive; second, impossible to clean; third, it requires great care in keeping the water above the tubes and crown-sheet. It is also very difficult to repair, and where the water is impure soon burns out.

Q. Are all steam-boilers fired alike?

A. No; some boilers, such as the locomotive, Dimfield, and Cornish boilers, are fired internally; while the cylinder, flue, and tubular boilers are fired externally; though in the last two the heat is applied both internally and externally.

MUD-DRUMS.

Q. What is a mud-drum?

A. A mud-drum is a small cylinder boiler, generally 24 inches in diameter, connected with the main boiler at the bottom, for the purpose of receiving the feed-water before it enters the boiler.

Q. What are the advantages and disadvantages of the mud-drum?

A. Mud-drums at one time were considered a benefit, as imparting extra heat, and also retaining the sediment carried in by the feed-water. But after years of experience in the use of the mud-drum, it was found that the drum imparted very little heat to the feed-water, and retained nothing but the earthy matter that is not injurious to a boiler. But all the destructive carbonates that form the hard glassy scale were carried into the boiler. Experience showed us that the mud-drum was positively dangerous, on account of not receiving sufficient heat to keep the iron dry. The corrosion of the mud-drum was very rapid, and became dangerous, in many cases before parties using them were actually aware of it. Some very serious accidents happened through the bursting of mud-drums. The mud-drum was generally worn out in 5 or 6 years, and the expense of removing the old drum and replacing it with a new one was more than the benefit derived from its use. The mud-drum is fast going out of use.

BOILER-HEADS.

Q. What kinds of material are most commonly used for boiler-heads ?

A. Wrought and cast iron.

Q. Which kind of material do you consider most safe ?

A. Wrought iron; as it is lighter, possesses greater strength, and presents better advantages for bracing in boilers of large diameter than cast iron.

Q. What are the most common shapes for cast-iron heads of boilers ?

A. Flat, concave, and convex.

Q. What are the advantages and disadvantages of the above-named heads?

A. First, the flat head should be avoided, like all flat surfaces in steam-boilers, because when subjected to a great strain they are positively dangerous. Second, the concave presents greater strength with less metal than the flat head. The only disadvantage we know of in the concave is in cases where the flange of the head is too deep on the inside, preventing the water from coming in contact with the rivets, and the boiler is liable to be burned through at that point. Third, the convex is not so strong as the concave, but is very much safer, with the same amount of metal, than the flat head.

BOILER-SHELLS.

Q. What thickness of boiler-iron do you consider the safest, most durable, and economical for boilers?

A. First, to insure safety in shells and flues of boilers, the thickness proper to use depends very much on the quality of the iron, diameter of boiler, and pressure to be carried. Second, as to durability, the thickest iron is not always the best, as the outside of the sheet becomes burned and crystallized, and in most cases gives less wear and satisfaction than a thinner gauge. Third, as to economy, thin boilers are more economical with fuel, and wear longer, provided in all cases that the diameter and the pressure are in proportion.

Q. What would you consider the proper thickness for all boilers?

A. The thickness of boiler-iron for all boilers must range between $\frac{7}{16}$ and $\frac{3}{16}$ of an inch, for the reason that ½ inch iron is almost too thick to be riveted and $\frac{3}{16}$ is too thin to be caulked. Of course, it will be understood that the choice of thickness between these two limits will depend on the quality of the iron, diameter of the boiler, and pressure to be carried, as before stated.

Q. What would you consider the proper thickness of iron for a boiler 48 inches in diameter, carrying a pressure of 90 pounds to the square inch?

A. ⅜ of an inch for ordinary brands of iron, but ¼ inch of superior iron would be just as strong.

Q. What do you consider the proper thickness for the wrought-iron heads of a boiler 48 inches in diameter, carrying a pressure of 90 pounds?

A. $\frac{11}{16}$ or ¾; either thickness would be perfectly safe.

Q. What would you consider the proper thickness for the tube-sheets and crown-sheets of boilers?

A. The thickness of the iron will depend on the diameter of the boiler and the quality of the iron, but the range for all boilers will be perhaps from ⅜ to ¾ of an inch thick.

Q. Is the temperature of thick boiler-plates often much higher than that of the steam?

A. Yes; thick boiler-plates or tube-sheets are in many instances several hundred degrees above the temperature of the steam in the boiler, and as a consequence are extremely dangerous.

Q. Is the pressure the same on all riveted seams in boiler-shells?

A. No; the pressure on the longitudinal rivets is nearly double that on the curvilinear rivets.

Q. What do you mean by longitudinal and curvilinear rivets?

A. By longitudinal rivets, I mean the rivets or seams that run lengthwise on the boiler; the curvilinear are those that are around the circumference of the shell.

Q. If the pressure on the longitudinal seams is double that on the curvilinear, how can all parts of the boiler sustain the same pressure?

A. By making the longitudinal seams double riveted and the curvilinear single.

Q. What is the difference in strength between single and double riveted seams?

A. Single-riveted seams are equal to 56 per cent. of the material used, while double riveting is equal to 70 per cent.

Q. What do you mean by "equal to 56 per cent. of material used"?

A. I mean that the boiler-plates lose 44 per cent. of their strength in the process of punching.

Q. What do you consider the proper diameter for rivets of boilers?

A. That would depend very much on the diameter of the boiler, thickness of iron, and pressure to be carried. For boilers from 36 to 42 inches diameter, and $\frac{3}{8}$ iron, if single riveted, the rivets ought to be $\frac{5}{8}$ of an inch for curvilinear, and $\frac{3}{4}$ for the longitudinal; if double riveted, $\frac{5}{8}$ will answer for both longitudinal and curvilinear seams. From $\frac{5}{16}$ iron down to $\frac{3}{16}$, smaller rivets will answer.

Q. Which do you consider the best method of riveting boilers — by hand or by machine?

A. For average or thin boiler-plates, hand riveting does very well, but for heavy iron, $\frac{7}{16}$ or $\frac{1}{2}$ inch thick, machine work is far superior, the power of the machine brings the work together far better and with less injury to the iron than can be done by hand.

Q. How should the fibre of the iron be placed to give the greatest strength?

A. The direction in which the iron is rolled should always be placed around the boiler, and not lengthwise, because in cylindrical boilers the strain in the line of the axis is much less than the circumferential bursting strain.

Q. Do you consider it an advantage to drill the rivet-holes in boilers instead of punching?

A. Yes; for all first-class work there can be no doubt but that all the rivet-holes ought to be drilled, on account of the liability of the plates to become fractured by the process of punching, causing a great reduction in the strength of the boilers.

Q. Would the same sized rivets answer for punched and drilled holes?

A. No; drilled holes require larger rivets than those that are punched.

Q. How do you explain that?

A. Because the shearing strain on the drilled holes is far greater than on the punched, as the hole is more perfect and sharp, and there is less friction between the sheets when drilled. A $\frac{5}{8}$ rivet will stand as much shearing strain in a punched hole as a $\frac{3}{4}$ will in a drilled hole.

Q. Do you consider the use of the drift-pin ought to be dispensed with as much as possible in making boilers?

A. Yes; a reckless use of the drift-pin has in many cases resulted in great injury to the boiler-plates; and there is good reason to believe that such injuries as are caused by the drift-pin, often hasten the destruction of the boiler.

Q. What is a drift-pin?

A. It is a tapering steel pin introduced into the holes in the seams, to bring them into line.

Q. How do you propose to dispense with the use of the drift-pin?

A. If the holes are laid off carefully in the sheet, and punched with judgment, there will be very little need for the drift-pin, as the holes can be straightened by a flat reamer. Such work will be greatly superior to that where the drift-pin is used.

Q. Do you think it would be any benefit to

slightly heat the boiler-plates before rolling them to form the shell of the boiler?

A. Yes; I think it would add very materially to the strength and durability of boilers if the sheets were rolled while warm, as the fibre of the iron would be drawn out; while, in the common practice of cold rolling, the fibre is crushed and broken.

Q. Does hammering improve the quality of iron?

A. No; it only hardens it, but at the same time renders it more brittle, whilst rolling imparts toughness.

Q. What fact is observable when boiler-iron is broken suddenly, as in the case of steam-boiler explosions?

A. It generally presents a crystalline fractured appearance; when, if broken by some slow process, it presents a fibrous or silky appearance, — in the first case the fibre is fractured, and in the other it is drawn out.

Q. What does the crystalline fracture indicate?

A. It indicates hardness, while a fibrous fracture is a mark of softness and ductility. The finer and more uniform the crystals the higher the quality of the iron.

Q. Is the pressure equal on all sides of the shell of a boiler when under steam?

A. No; there is more pressure on the lower than on the upper side of a boiler; as the steam presses equally on the surface of the water as on the upper side of the boiler, the weight of the water must be added to the pressure on the lower side.

Q. How do you explain that?

A. In this way: in a boiler 36 inches in diameter, circumference 113 inches, if full of water, the water would weigh 441 pounds to the foot in length; now, if the boiler was 20 feet long, the water would weigh 8820 pounds; take ⅔ of that (which would be about the quantity of water in a boiler when under steam) and the weight of water would be 5880 pounds, which shows the excess of pressure on the lower side of the boiler.

Q. Are the shells and flues of boilers sometimes injured by the application of the cold water or "Hydrostatic" test?

A. Yes; the shells and flues of boilers are sometimes injured by a reckless use of the test, and in many cases explosions take place soon after the test is applied.

Q. Would the shell and flues of a boiler be stronger under a cold water pressure of 70 or 80 pounds to the square inch than under the same steam pressure?

A. No; as iron increases in strength by the application of heat up to 550° Fah., the boiler would be stronger under the steam pressure.

Q. What is the maximum strength of wrought iron?

A. 60,000 pounds to the square inch at a temperature of 550° Fah., or in other words, a bar of iron one inch square would stand a tensile strain of 60,000 pounds.

A Table deduced from Experiments on Iron Plates for Steam-Boilers, by the Franklin Institute, Philadelphia.

Iron boiler-plate was found to increase in tenacity as its temperature was raised, until it reached a temperature of 550° above the freezing-point, at which point its tenacity began to diminish. The following table exhibits the cohesive strength at different temperatures: —

At	32° to 80°	tenacity	is	56,000 lbs.	or one-seventh below its maximum.
At	570°	"	"	66,000 "	the maximum.
At	720°	"	"	55,000 "	the same nearly as at 32°.
At	1050°	"	"	32,000 "	nearly one-half the maximum.
At	1240°	"	"	22,000 "	nearly one-third the maximum.
At	1317°	"	"	9,000 "	nearly one-seventh the maximum.

At 3000° wrought iron becomes fluid.

It will be seen that by the accidental overheating of a boiler the tenacity of the iron would soon become reduced to nearly half its strength.

The following Table is the result of experiments made on different brands of boiler-iron at the Stevens Institute of Technology. The samples tested were taken from lots of iron sent to market by various manufacturers of boiler-plate, and may be considered to represent in most cases an average quality of the several grades of iron.

Thirty-three experiments were made upon iron taken from the exploded steam-boiler of the ferry-boat Westfield. The following were the results:

	Lbs. per sq. inch
Average breaking weight	41,653
16 experiments made upon high grades of American boiler-plate.	
Average breaking weight	54,123
15 experiments made upon high grades of American flange iron.	
Average breaking weight	42,144
6 experiments made upon English Bessemer steel.	
Average breaking weight	82,621
5 experiments made upon best English Lowmoor boiler-plate.	
Average breaking weight	58,984
6 experiments made upon samples of tank iron from different manufacturers.	
Average breaking weight, No. 1	43,831
" " " No. 2	42,011
" " " No. 3	41,249
2 experiments made on iron taken from the exploded steam-boiler of the Red Jacket.	
Average breaking weight	49,000

It will be noticed that the above experiments reveal a great variation in the strength of boiler-plate of different grades of iron, and furnish conclusive evidence that the tensile strength of boiler-iron ought to be taken at 50,000 pounds to the square inch, instead of 60,000.

Q. What reduction in length does wrought iron undergo when subjected to compression?

A. It is reduced .0001 of its length by every ton

per square inch up to 13 tons, beyond which the amount of compression increases more rapidly.

STEEL BOILERS.

In stiffness or rigidity, as well as high ultimate resistance and good conducting powers, steel presents many advantages, as a substitute for iron, in the construction of boilers. Some of the largest locomotive establishments in this country are using steel for boilers. The substitution of steel for iron in the manufacture of boilers is rapidly spreading in all the countries of Europe.

Q. What advantages do steel boilers possess over iron?

A. The advantages of steel boilers are: first, they are lighter; second, they are stronger; third, their conducting powers are higher. For instance, a steel boiler 48 inches in diameter and $\frac{1}{4}$ inch thickness of sheets is as strong as an iron boiler of the same diameter and $\frac{3}{8}$ inch thickness of iron, and will evaporate 25 per cent. more water with the same amount of fuel in a given time. Steel boilers are more free from incrustation and corrosion than iron, on account of the density and compactness of the material. But in the manufacture of steel boilers great care should be taken that the plates are not fractured in punching.

INTERNAL AND EXTERNAL PRESSURES.

An important difference of condition, never to be lost sight of in boiler-work, is that all internal pressure tends to preserve the symmetry of form of the boiler, while all external pressure tends to destroy that symmetry, which, if altered at all, places the boiler in new conditions, likely to end in sudden destruction.

Q. If pressure is exerted on the internal or external surface of the cylinder, is the effect the same in both cases?

A. No; when pressure is exerted within a tube or cylinder, the tendency of the strain is to cause the tube to assume the true cylindrical form; but when pressure is exerted on the outside of the tube, the tendency of that pressure is to crush the tube or flatten it; as it is a well-known fact that iron of any strength when formed into a tube will require a much greater strain to tear it asunder than it would take to crush it.

Q. How do you explain that?

A. In this way: a thin hoop of iron will resist a very great amount of tearing force, but if that same hoop or circle be placed as a prop under half the weight that was exerted to tear it apart, it would be crushed flat.

Q. What is the difference between external and internal strain?

A. Internal is a tearing strain, whilst external is a crushing strain; and flues and tubes of boilers are nothing but a series of props, and a constant tendency of the pressure is to flatten the tube or flue, and cause it to collapse.

Q. What is a collapse?

A. It is the crushing or flattening of a flue by over-pressure, and is often attended with terrible results.

Q. What is an explosion?

A. It is an accident that takes place in most cases from the water becoming low, when, through ignorance or recklessness, cold water is admitted to the boiler, and explosion takes place.

Q. What is a burst?

A. It is an accident that occurs from over-pressure, and is generally more terrible in effect than either an explosion or a collapse.

Q. Why should the burst be more destructive than the explosion?

A. Because, in the case of the explosion, the tenacity of the iron is partly destroyed by being over-heated; while, in the case of the burst, the iron is torn apart at its maximum strength.

Q. What is the best preventive against explosions, bursts and collapses?

A. Never tax the boiler above a safe working pressure; keep it clean and in good repair, the safety-valve in good order, and the water at all times at the proper level.

Q. What part gives way first when a boiler bursts?

A. A boiler gives way in the weakest place.

Q. What should be considered the strength of a boiler?

A. The weakest part of the boiler.

Q. How do you explain that?

A. In this way: if some parts of a boiler are $\frac{3}{8}$ of an inch thick, whilst other parts are not more than $\frac{3}{6}$, the pressure carried on the boiler should be that which would be safe for $\frac{3}{16}$ of an inch.

RULES.

Rule for finding the quantity of water in a boiler in cubic inches.

Multiply the internal area of the head of the boiler, in inches, by the length of the boiler in inches, the product will be the number of cubic inches of water in the boiler. Divide this product by 1728, and the quotient will be the number of cubic feet of water in the boiler.

Q. In what way do riveted joints give way?

A. Either by shearing off the rivets in the middle of their length, or by tearing through one of the plates in the line of the rivet-holes. In a perfect joint, the rivets should be on the point of shearing just as the plates were about to tear; hence, in good practice, it is an established rule to employ a certain number of rivets to the linear foot. If these are placed in a single row, the rivet-holes so nearly approach each other that the strength of the plates is much reduced; but if they are arranged in two lines, a greater number may be used, and yet more space left between the holes, and greater strength and stiffness imparted to the plates at the joint.

It is perfectly obvious that in perforating a line of holes along the edge of a plate, we must greatly reduce its strength.

When this great deterioration of strength at the joint is taken into account, it cannot but be of the greatest importance that in structures subjected to such violent strains as steam-boilers are, the strongest method of riveting should be adopted.

Rule for finding safe external pressure on boiler-flues.

Multiply the square of the thickness of the iron by the constant whole number 806,300; divide this product by the diameter of the flue in inches; multiply the quotient by the length of the flue in feet; multiply this product by 3.

Rule for finding safe working pressure of any boiler.

Multiply the thickness of the iron by .56 (or .70, according as the boiler is single or double riveted); multiply this product by 10,000 (safe load); then divide this last product by the internal radius: the quotient will be the safe working pressure in pounds per square inch.

Five times the safe working pressure is the bursting pressure.

The experiments of Mr. Fairbairn and others have established the following relative strengths as the value of plates with either single or double riveted joint.

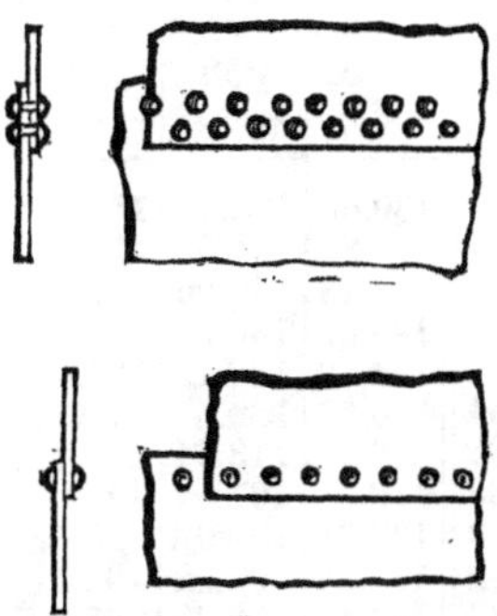

Taking the strength of the plate at	100
The strength of the double-riveted joint would then be . . .	70
And the strength of the single-riveted joint	56

The following Table will show the safe working pressure for boilers of different diameters and different thicknesses of iron, according to Fairbairn's experiments.

For internal pressure, $\frac{1}{5}$ of the value of ordinary boiler-iron, or 50,000 pounds for the inch of section, is taken to be safe, while in external pressures $\frac{1}{3}$ is taken.

TABLE.

INTERNAL PRESSURES.

Birmingham Wire Gauge.		3/8	00	0	1	2
Thickness of Iron.		.375 3/8	.358 3/8 Scant.	.340 11/32	.300 5/16	.284 9/32
	In.					
External	24	180.65	172.20	163.29	143.59	135.75
Diameter.	26	166.34	158.58	150.39	132.28	125.08
	28	154.13	146.96	139.38	122.63	115.95
	30	143.59	136.92	129.88	114.29	108.07
	32	134.40	128.17	121.58	107.01	101.20
	34	126.31	120.47	114.29	100.60	95.14
	36	119.15	113.64	107.81	94.92	89.77
Longitudinal	38	112.75	107.54	102.04	89.84	84.98
Seams,	40	107.01	102.07	96.85	85.28	80.67
Single	42	101.81	97.12	92.11	81.16	76.77
Riveted.	44	97.11	92.63	87.90	77.42	73.24
	46	92.82	88.54	84.02	74.01	70.01
	48	88.89	84.80	80.47	70.89	67.06
	50	85.28	81.36	77.21	68.02	64.35
	52	81.95	78.18	74.20	65.37	61.84
	54	78.87	75.25	71.42	62.92	59.53
	56	76.02	72.53	68.84	60.65	57.38
	58	73.36	70.00	66.43	58.54	55.38
	60	70.89	67.63	64.19	56.57	53.52
	62	68.57	65.43	62.10	54.72	51.78
	64	66.40	63.36	60.14	53.00	50.15
	66	64.37	61.42	58.30	51.38	48.61
	68	62.45	59.59	56.57	49.85	47.17
	70	60.65	57.87	54.93	48.41	45.81
	72	58.95	56.25	53.39	47.06	44.53
	74	57.34	54.71	51.94	45.78	43.32
	76	55.81	53.26	50.56	44.56	42.17
	78	54.37	51.88	49.25	43.41	41.08
	80	53.00	50.57	48.01	42.32	40.04

TABLE.

INTERNAL PRESSURES.—(Continued.)

Birmingham Wire Gauge.		3	4	5	6	7	8
Thickness of Iron.		.259 $\frac{1}{4}$ Full.	.238 $\frac{1}{4}$ Scant.	.220 $\frac{7}{32}$	.203 $\frac{6}{23}$ Full.	.180 $\frac{6}{32}$ Sc't.	.165 $\frac{5}{32}$ Full.
	In.						
External	24	123.53	113.31	104.58	96.36	85.28	78.07
Diameter	26	113.84	104.44	96.40	88.83	78.63	71.99
	28	105.55	96.85	89.40	82.39	72.94	66.79
	30	98.39	90.29	83.36	76.83	68.02	62.29
	32	92.14	84.56	78.07	71.96	63.72	58.35
	34	86.64	79.51	73.42	67.68	59.93	54.89
	36	81.75	75.04	69.29	63.88	56.57	51.81
Long.	38	77.39	71.04	65.60	60.48	53.56	49.06
Seams.	40	73.47	67.44	62.29	57.42	50.86	46.58
Single	42	69.93	64.19	59.29	54.66	48.41	44.35
Riveted.	44	66.71	61.24	56.57	52.15	46.20	42.32
	46	63.78	58.55	54.08	49.87	44.17	40.46
	48	61.09	56.09	51.81	47.77	42.32	38.77
	50	58.62	53.82	49.72	45.84	40.61	37.21
	52	56.35	51.74	47.79	44.07	39.04	35.77
	54	54.24	49.80	46.00	42.42	37.58	34.43
	56	52.28	48.01	44.35	40.90	36.23	33.20
	58	50.46	46.34	42.81	39.48	34.98	32.04
	60	48.77	44.78	41.37	38.15	33.80	30.97
	62	47.18	43.33	40.03	36.91	32.71	29.97
	64	45.69	41.96	38.77	35.75	31.68	29.02
	66	44.30	40.68	37.58	34.66	30.71	28.14
	68	42.99	39.48	36.47	33.64	29.80	27.31
	70	41.75	38.34	35.42	32.67	28.95	26.53
	72	40.58	37.27	34.43	31.76	28.14	25.78
	74	39.48	36.25	33.50	30.89	27.38	25.08
	76	38.43	35.29	32.61	30.08	26.65	24.42
	78	37.44	34.38	31.77	29.30	25.96	23.79
	80	36.49	33.52	30.97	28.56	25.31	23.20

TABLE.

INTERNAL PRESSURES. — (Continued.)

Birmingham Wire Gauge.		3/8	00	0	1	2
Thickness of Iron.		.375 3/8	.358 3/8 Scant.	.340 11/32	.300 5/16	.284 9/32
	In.					
External	24	225.81	215.26	204.12	179.49	169.67
Diameter.	26	207.93	198.23	187.91	165.35	156.34
	28	192.66	183.70	174.23	153.28	144.94
	30	179.49	171.15	162.35	142.86	135.09
	32	168.00	160.21	151.98	133.76	126.49
Longitudinal	34	157.89	150.58	142.86	125.75	118.93
Seams,	36	148.94	142.05	134.77	118.64	112.21
Double	38	140.94	134.43	127.55	112.30	106.22
Riveted.	40	133.76	127.58	121.06	106.60	100.83
Curvilinear	42	127.27	121.40	115.20	101.45	95.96
Seams,	44	121.39	115.79	109.88	96.77	91.55
Single	46	116.02	110.68	105.03	92.51	87.52
Riveted.	48	111.11	106.00	100.59	88.61	83.83
	50	106.19	101.70	96.51	85.02	80.43
	52	102.44	97.73	92.75	81.71	77.31
	54	98.59	94.10	89.27	78.69	74.41
	56	95.02	90.66	86.04	75.81	71.73
	58	91.70	87.49	83.04	73.17	69.23
	60	88.61	84.54	80.24	70.71	66.90
	62	85.71	81.78	77.63	68.40	64.72
	64	83.00	79.17	75.17	66.25	62.68
	66	80.46	76.78	72.87	64.22	60.77
	68	78.07	74.49	70.71	62.31	58.96
	70	75.81	72.34	68.67	60.52	57.26
	72	73.68	70.31	66.74	58.82	55.66
	74	71.67	68.39	64.92	57.22	54.15
	76	69.77	66.60	63.19	55.70	52.71
	78	67.96	64.85	61.56	54.26	51.35
	80	66.25	63.22	60.01	52.90	50.06

TABLE.

INTERNAL PRESSURES. — (Continued.)

Birmingham Wire Gauge.		3	4	5	6	7	8
Thickness of Iron.		.259 ¼ Full.	.238 ¼ Scant.	.220 7/32	.203 6/32 Full.	.180 6/32 Scant.	.165 5/32 Full.
	In.						
Ext'l	24	154.42	141.64	130.73	120.45	106.60	97.59
Diam.	26	142.30	130.54	120.50	111.04	98.21	89.99
	28	131.94	121.06	111.76	102.99	91.17	83.48
	30	122.99	112.86	104.19	96.03	85.02	77.86
	32	116.32	105.70	97.59	89.95	79.65	72.94
Long.	34	108.30	99.39	91.78	84.60	74.91	68.61
Seams,	36	102.19	93.80	86.61	79.84	70.71	64.76
Doub'e	38	96.74	88.80	82.00	75.60	66.95	61.32
Riv't'd	40	91.84	84.30	77.86	71.78	63.57	58.23
Curvil.	42	87.41	80.24	74.11	68.33	60.52	55.44
Seams,	44	83.39	76.56	70.71	65.19	57.75	52.90
Single	46	79.72	73.19	67.60	62.33	55.22	50.58
Riv't'd	48	76.37	70.11	64.76	59.71	52.90	48.46
	50	73.28	67.28	62.11	57.31	50.77	46.51
	52	70.43	64.67	59.74	55.08	48.80	44.71
	54	67.80	62.25	57.51	53.40	46.98	43.04
	56	65.35	60.01	55.44	51.12	45.29	41.50
	58	63.07	57.92	53.51	49.35	43.72	40.06
	60	60.96	55.98	51.71	47.69	42.25	38.71
	62	58.98	54.16	50.03	46.14	40.88	37.46
	64	57.12	52.45	48.46	44.69	39.60	36.28
	66	55.37	50.85	46.98	43.33	38.39	35.18
	68	53.73	49.35	45.59	42.05	37.26	34.14
	70	52.19	47.93	44.28	40.84	36.19	33.16
	72	50.73	46.59	43.04	39.70	35.18	32.23
	74	49.35	45.32	41.87	38.62	34.22	31.36
	76	48.04	44.11	40.76	37.60	33.32	30.53
	78	46.80	42.98	39.71	36.63	32.46	29.74
	80	45.62	41.90	38.71	35.71	31.64	28.99

FOAMING IN STEAM-BOILERS.

Q. What is the cause of foaming in steam-boilers?

A. Foaming in steam-boilers might be attributed to different causes. First, to the boiler not having a sufficient amount of steam-room, so that whenever the valve opens to admit steam to the cylinder, the pressure on the surface of the water is lessened, allowing the water to rise up from the bottom of the boiler. Second, foaming is sometimes caused by the foul condition of the boiler; but in such cases it will be easy to discover the cause, as the water in the glass gauge will appear quite muddy. Third, foaming is caused by the presence of any substance of a soapy or greasy nature in the water. But whatever may be the cause of foaming, it is always attended with great danger to the boiler and a certain amount of injury to the engine.

In all cases where a boiler foams badly, the water is lifted from the fire-surface of the boiler, and allows the iron to burn; also, the mud and water from the boiler are carried over with the steam to the cylinder, occupying the clearance between the piston and the head of the cylinder; not only destroying the surface of the cylinder by the grit and dirt, but in many cases causing the fracture of the cylinder-head.

Q. What is the best preventive against foaming?

A. The best preventive against foaming is: First, a clean boiler. Second, pure water. Third, a sufficient amount of steam-room. Fourth, a steam-pipe well proportioned to the size of the cylinder.

RUST.

Q. From what cause does the decay of iron arise?

A. From the joint action of air and water, the oxygen from which combines with the iron and forms a hydrate called rust.

Q. What parts of steam-boilers are most liable to corrode or burn out?

A. The parts directly over the fire when coated with scale, and those exposed to dampness.

PATENT STEAM-BOILERS.

There is hardly anything connected with steam-engineering that is so uncertain as the records of the results of experiments made with patent boilers. A great many new designs have been brought forward and tested, but the results have been either imperfectly recorded or kept secret by men who did not care to announce that they had been unsuccessful. It is to be regretted that steam users and the public know so little of the evaporative capacity of patent boilers; even such information as we do have is very imperfect, as those who supply it are almost invariably inaccurate. First, because they do not know how to secure accuracy; second, because they deceive themselves; and, third, because they deceive the public.

There are three great objects to be attained in designing a steam-boiler; in the first place it should be safe; in the second, it should be economical with fuel; in the third, it should steam well. Now, as regards patent boilers, there is information wanted concerning them on the three above points.

THE SAFETY-VALVE.

THE form and construction of this indispensable adjunct to the steam-boiler are of the highest importance, not only for the preservation of life and property, which would in the absence of this means of safety be constantly jeopardized, but also to secure the durability of the steam-boiler itself.

Increasing the pressure to a dangerous degree would be impossible in any boiler, if the safety-valve were what it is supposed to be,—a perfect means for liberating all the steam which a boiler may produce with the fires in full blast, and all other means for the escape of steam closed. Until such a safety-valve shall be devised and adopted into general use, safety from gradually increasing pressure must depend on the attention and watchfulness of the engineer.

We have decidedly too much theory on the safety-valve, and most of this theory is the merest vagary, which it is impossible to harmonize with experience and sound practice. All that the safety-valve needs to make it what it was intended to be, is, first, an orifice proportioned to the grate-surface; second, simplicity of construction; third, directness of action.

Q. What is the object or use of the safety-valve?

A. It is a valve intended to relieve the boiler from extra pressure, and prevent bursting, collapse or explosion.

Q. What do you consider a proper proportion for the safety-valve of a boiler?

A. The area of the safety-valve should be ½ square inch to each square foot of grate-surface.

Q. Will this amount of opening of safety-valve be safe for any ordinary pressure?

A. Yes; it will be safe for any pressure from 10 pounds to the square inch up to 100 pounds.

Q. Is an enlargement of the safety-valve greatly beyond what is customary in common practice, dangerous?

A. Yes; if such a safety-valve by any accident should be knocked or lifted suddenly from its seat, it would probably cause the destruction of the boiler.

Q. Should every steam-boiler have two safety-valves?

A. No; one safety-valve of suitable proportions, and in good order, is sufficient.

Q. How should the safety-valve be kept or cared for?

A. It should always be kept as free as possible from dust and ashes, and all its working parts in good order.

Q. How often should the safety-valve be moved?

A. At least once a day, more particularly in the morning.

Q. Why should the safety-valve be moved in the morning?

A. So as to be sure that it is in good working order before starting the fire.

Q. What are the most important principles to be adhered to in the construction of the safety-valve?

A. Simplicity of construction, directness and freedom of action.

Q. Does the safety-valve become worn and leaky by the continual action of the steam?

A. Yes; all safety-valves become leaky, and ought to be ground carefully on their seats.

Q. What is the best material to use for grinding safety-valves?

A. Pulverized glass, grit of grinding-stones, or fine emery.

Q. Should safety-valves be constructed with loose or vibratory stems?

A. Yes; as the rigid or solid stem is apt to become jammed by the canting of the lever and weight, and in such cases the higher the pressure the more difficult is the action of the valve.

Q. Is the marking on safety-valves sometimes incorrect?

A. Yes; decidedly so.

Q. How can you tell whether the safety-valve lever is marked correctly or not?

A. By calculation.

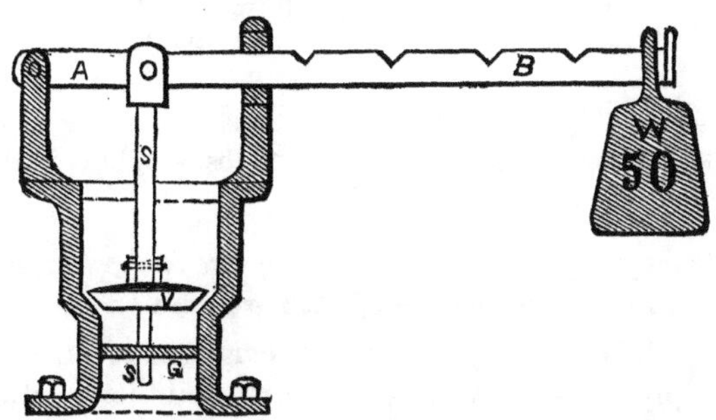

B, lever. A, short arm of lever. S S, stem. V, valve. G, guide. W. weight.

RULES.

Rule for finding the weight necessary to put on a lever, when the area of valve, pressure, etc., are known.

Multiply the area of valve by the pressure in pounds per square inch; multiply this product by the distance of the valve from the fulcrum; multiply the weight of lever by one-half its length (or its centre of gravity); then multiply the weight of valve and stem by their distance from the fulcrum; add these last two products together, and subtract their sum from the first product, and divide the remainder by the length of lever; the quotient will be the weight of the ball.

EXAMPLE.

Area of valve,	7 sq. in.	60 lbs.	9 lbs.	6 lbs.
Pressure,	60 lbs.	7 in.	12 in.	3 in.
		420 lbs.	108 lbs.	18 lbs.
Fulcrum,	3 in.	3 in.	18 lbs.	
		1260 lbs.	126 lbs.	
		126 lbs.		
Length of lever,	24 in.	24)1134 lbs.		
Weight of lever,	9 lbs.	47.25 lbs. weight of ball.		
Weight of valve and stem,	6 lbs.			

Rule for finding the Pressure per Square Inch when the Area of Value, Weight of Ball, etc., are known.

Multiply the weight of ball by length of lever, and multiply the weight of lever by one-half its length, (or its centre of gravity;) then multiply the weight of valve and stem by its fulcrum. Add these three products together. This sum, divided by the product of the area of valve, and its distance from the fulcrum, will give pressure in pounds per square inch.

EXAMPLE.

Area of valve,	9 sq. in.	60 lbs.	20 lbs.	5 lbs.
Fulcrum,	4 in.	36 in.	18 in.	4 in.
Length of lever,	36 in.	2160 lbs.	360 lbs.	20 lbs.
Weight of lever,	20 lbs.	360 lbs.		
Weight of ball,	60 lbs.	20 lbs.	9 in.	
Weight of valve and stem,	5 lbs.	36)2580 lbs.	4 in.	
		70.55 lbs.	36 in.	
		Pressure per sq. in.		

Rule for finding the Area of a Circle when the Diameter is given.

Multiply the square of the diameter by the decimal .7854.

EXAMPLE. Diameter, 3 inches,

3	.7854
3	9
9 sq. in.	70.686 sq. in. = area.

Q. How do you square a diameter?

A. Any diameter multiplied by itself is squared; as, for instance, 10 squared equals 100.

Q. Why do you multiply the square by .7854?

A. By squaring the diameter we get square inches, and if we multiply by .7854, we get circular inches.

Q. What is the difference between circular and square inches?

A. A circular inch is .7854 part of the square inch.

Q. What do you mean by the word "area"?

A. By *area* we mean the amount of surface exposed to the action of the steam.

A TABLE FOR SAFETY-VALVES.

Containing the Circumferences and Areas of Circles from $\frac{1}{16}$ of an inch to 4 inches.

Diameter.	Circumference.	Area.	Diameter.	Circumference.	Area.
$\frac{1}{16}$	.1963	.0030	2 in.	6.2832	3.1416
$\frac{1}{8}$	.3927	.0122	$\frac{1}{16}$	6.4795	3.3411
$\frac{3}{16}$	.5890	.0276	$\frac{1}{8}$	6.6759	3.5465
$\frac{1}{4}$	.7854	.0490	$\frac{3}{16}$	6.8722	3.7582
$\frac{5}{16}$	.9817	.0767	$\frac{1}{4}$	7.0686	3.9760
$\frac{3}{8}$	1.1781	.1104	$\frac{5}{16}$	7.2649	4.2001
$\frac{7}{16}$	1.3744	.1503	$\frac{3}{8}$	7.4613	4.4302
$\frac{1}{2}$	1.5708	.1963	$\frac{7}{16}$	7.6576	4.6664
$\frac{9}{16}$	1.7671	.2485	$\frac{1}{2}$	7.8540	4.9087
$\frac{5}{8}$	1.9635	.3068	$\frac{9}{16}$	8.0503	5.1573
$\frac{11}{16}$	2.1598	.3712	$\frac{5}{8}$	8.2467	5.4119
$\frac{3}{4}$	2.3562	.4417	$\frac{11}{16}$	8.4430	5.6727
$\frac{13}{16}$	2.5525	.5185	$\frac{3}{4}$	8.6394	5.9395
$\frac{7}{8}$	2.7489	.6013	$\frac{13}{16}$	8.8357	6.2126
$\frac{15}{16}$	2.9452	.6903	$\frac{7}{8}$	9.0321	6.4918
			$\frac{15}{16}$	9.2284	6.7772
1 in.	3.1416	.7854			
$\frac{1}{16}$	3.3379	.8861	3 in.	9.4248	7.0686
$\frac{1}{8}$	3.5343	.9940	$\frac{1}{16}$	9.6211	7.3662
$\frac{3}{16}$	3.7306	1.1075	$\frac{1}{8}$	9.8175	7.6699
$\frac{1}{4}$	3.9270	1.2271	$\frac{3}{16}$	10.0138	7.9798
$\frac{5}{16}$	4.1233	1.3529	$\frac{1}{4}$	10.2102	8.2957
$\frac{3}{8}$	4.3197	1.4848	$\frac{5}{16}$	10.4065	8.6179
$\frac{7}{16}$	4.5160	1.6229	$\frac{3}{8}$	10.6029	8.9462
$\frac{1}{2}$	4.7124	1.7671	$\frac{7}{16}$	10.7992	9.2806
$\frac{9}{16}$	4.9087	1.9175	$\frac{1}{2}$	10.9956	9.6211
$\frac{5}{8}$	5.1051	2.0739	$\frac{9}{16}$	11.1919	9.9678
$\frac{11}{16}$	5.3015	2.2365	$\frac{5}{8}$	11.3883	10.3206
$\frac{3}{4}$	5.4978	2.4052	$\frac{11}{16}$	11.5846	10.6796
$\frac{13}{16}$	5.6941	2.5801	$\frac{3}{4}$	11.7810	11.0446
$\frac{7}{8}$	5.8905	2.7611	$\frac{13}{16}$	11.9773	11.4159
$\frac{15}{16}$	6.0868	2.9483	$\frac{7}{8}$	12.1737	11.7932
			$\frac{15}{16}$	12.3700	12.1768
			4 in.	12.5664	12.5664

For larger Areas, see Table for Cylinders, page 183.

FEED-WATER HEATERS.

THE waste of fuel and danger arising from the rapid incrustation of steam-boilers from the deposit of mineral substances of impure water are well known; and the damage done to sheets and rivets by overheating, in consequence of heavy deposits of mineral substances, is one of the causes which produce explosions in steam-boilers, causing, as they do, terrible destruction and loss of life. He whose inventive genius and mechanical skill contribute to avert such disasters by purifying the feed-water for steam-boilers, will, in point of safety and economy, confer a boon on mankind.

The saving in fuel that might be effected by thoroughly heating the feed-water — by means of the exhaust-steam in a properly constructed heater — would be immense, which will be seen from the following facts: —

A pound of feed-water entering a steam-boiler at a temperature of 50° Fah., and evaporating into steam of 60 pounds pressure, requires as much heat as would raise 1157 pounds of water 1 degree. A pound of feed-water raised from 50° Fah. to 220 Fah., requires 987 thermal units of heat; which, if absorbed from exhaust-steam passing through a heater, would be a saving of 15 per cent. in fuel. Feed-water, at a temperature of 200° Fah., entering a boiler, as compared in point of economy with feed-water at 50° Fah., would effect a saving of over 13 per cent. in fuel; and with a well-constructed heater there ought to be no trouble in raising the feed-water to a temperature of 212° Fah.

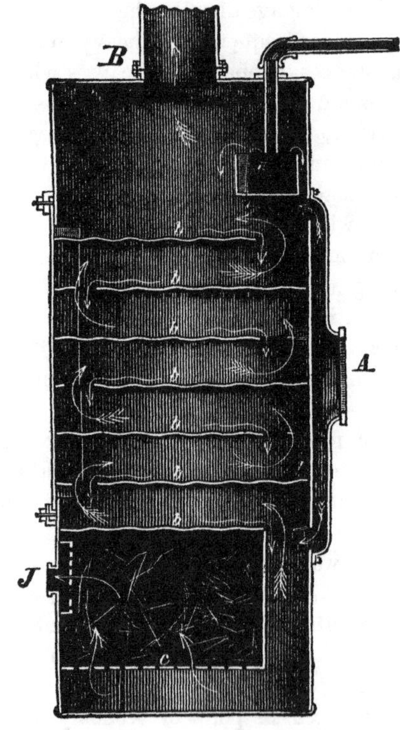

F

STILWELL'S PATENT IMPROVED LIME-EXTRACTING HEATER AND FILTER COMBINED.

Has obtained a world-wide reputation as the only lime-extracting heater extant that will positively and entirely prevent incrustation in steam-boilers. It is indispensable to an economical use of steam, and does its work in the only safe and rational manner, by removing all impurities from the feed-water before it enters the boilers.

EXPLANATION OF CUT.

A.—Steam enters the heater, and is divided into two currents. *B.*—Steam escapes from the heater. *J.*—Hot water leaves heater. *a.*—Overflow-cup suspended on end of cold-water pipe. *bbbb.*—Removable shelves or depositing surfaces. *c.*—Filtering chamber, to be filled with any suitable filtering material. The feathered arrows indicate the course of the steam, and the plain arrows the course of the water.

The device is claimed to be an established success, over 3000 being now at work. We are informed that it has been fully tested over a period of 9 years, and that it is guaranteed by its makers to completely prevent incrustation.

They are manufactured by the Stilwell & Bierce Manufacturing Company, Dayton, Ohio.

FUEL.

Q. What are the constituents of coal?

A. The chief constituent is carbon.

Q. How much carbon does good coal contain?

A. About 90 per cent., and 10 per cent. of earthy matter.

Q. Are there any other gases in coal except carbon?

A. Yes; hydrogen, nitrogen, and sulphur in small quantities.

Q. How much heat does 1 pound of pure carbon yield in burning?

A. 14,000 units.

Q. How much heat does 1 pound of good coal, containing 90 per cent. of carbon, produce?

A. It produces in burning about 13,000 units of heat.

Q. What is the mechanical equivalent of 13,000 units?

A. 10,036,000 pounds; that is to say, 10,036,000 pounds raised 1 foot high.

Q. How much air does it require to burn 1 pound of coal?

A. 240 cubic feet.

Q. How much air does it require to burn 100 pounds of coal?

A. 15,524 cubic feet of air.

Q. What is the difference between anthracite and bituminous coal?

A. Anthracite coal consists of about 90 per cent. of carbon, and bituminous about 80 per cent.

Q. What is the difference between anthracite coal and good pine wood?

A. 1 pound of good anthracite coal will evaporate as much water as 2½ pounds of wood.

Q. What is the difference between anthracite coal and coke?

A. Good coke will evaporate more water than the best anthracite coal.

FIRE.

Q. What is fire?

A. It is the rapid combustion of the constituent elements of organic matter.

Q. What are the causes which originate this rapid decomposition of organic matter?

A. Fire sometimes occurs without any visible cause when materials are in a favorable position, and under such circumstances is called spontaneous combustion.

Q. Is there any other cause from which fire originates?

A. Yes; fire is sometimes induced by friction, sometimes kindled by electricity; but the mode in which it is most commonly propagated is from the combustion of other matter that has been already ignited.

TABLE.

Showing the Total Heat of Combustion of Various Fuels.

SORT OF FUEL.	Equivalent in pure carbon.	Evaporative power in lbs. water from 212° Fah.	Total heat of combustion in lbs. water heated 1° Fah.
Charcoal	0.93	14.00	13,500
Charred peat	0.80	12.00	11,600
Coke — good	0.94	14.00	13,620
" mean	0.88	13.20	12,760
" bad	0.82	12.30	11,890
COAL:			
Anthracite	1.05	15.75	15,225
Hard bituminous — hardest	1.06	15.90	15,370
" " softest	0.95	15.25	13,775
Coking coal	1.07	16.00	15,837
Cannel coal	1.04	15.60	15,080
Long flaming splint coal	0.91	13.65	13,195
Lignite	0.81	12.15	11,745
PEAT:			
Perfectly air-dry	0.66	10.00	9,660
Containing 25 per cent. water	......	7.75	7,000
WOOD:			
Perfectly air-dry	0.50	7.50	7,245
Containing 25 per cent. water	......	5.80	5,600

REMARK. — In a boiler of fair construction, a pound of coal will convert 9 pounds of water into steam. Each pound of this steam will represent an amount of energy, or capacity for performing work, equivalent to 746,666 footpounds, or for the whole 9 pounds, 6,720,000 footpounds. In other words, 1 pound of coal has done as much work in evaporating 9 pounds of water into 9 pounds of steam as would lift 2,232 tons 10 feet high.

CHIMNEYS.

PROBABLY there is nothing connected with the generation of steam and utilization of heat so imperfectly understood, at present, as the quantity of air that passes through the furnaces of boilers under varying conditions of draught; it has been generally assumed, from experiment in common practice, that double the amount of air necessary for complete combustion of the fuel passes through the furnace.

Experiments are greatly needed to determine the state of combustion in varying dimensions of chimneys, as well as the quantity of air drawn through the furnaces under these varying rates of combustion.

Q. What is a chimney?

A. It is the machine or opening through which the air and smoke from the furnace pass.

Q. Is there any certain rule laid down for the height of chimneys?

A. No, as much depends on the position; if a stack or chimney is surrounded by hills or high buildings, it ought to be higher than those in more exposed places.

Q. Is it necessary to have higher chimneys for long boilers than for short ones?

A. Yes; chimneys should be higher for long boilers than for short ones, but it often happens that short chimneys draw better than long ones.

Q. Does it make any difference whether the inside of the chimney is round or square?

A. Yes; as round chimneys have invariably more draft than square ones.

Q. Does it make any difference whether the opening in a chimney is more or less than the area of the flues or tubes?

A. Yes; the opening in the chimney ought to be in all cases one-fifth greater than the combined areas of the flues or tubes; if it is less, it will destroy the draught.

Q. Should the opening in the chimney be wider at the bottom than at the top?

A. No; it should be just the reverse: it should be wider at the top than at the bottom; if it is not, it will retard the draught.

SMOKE.

When coal is burned in an open furnace the principal products of combustion are carbonic acid and water; certain portions of carbon escape combustion and constitute what is commonly called smoke, which is about 25 per cent. of the fuel in the shape of unconsumed gases and vapors.

Various methods have been resorted to in this country and England for the purpose of burning smoke, but with only partial success.

Q. Will you explain some of the different arrangements brought forward at different times for the purpose of burning smoke?

A. James Watt used a hopper to feed the coal into the furnace, supposing that the small quantity of gas or smoke evolved by that process would be readily consumed, but the plan soon fell into disuse.

Another device was to place a small furnace behind the bridge wall of the boiler, so that the smoke and gases passing over it might be ignited and consumed, but this was soon abandoned. Another method was to supply coal to one side of the furnace at a time, in the hope that the gas and smoke from the fresh coal would pass over the clean fire on the other side and be consumed; but the carrying out of this plan involved great difficulties, and had to be abandoned. Another attempt to burn smoke was by introducing air into the furnace above the fire, through small holes in the door; but the attempt was attended with no better success than the former, as the quantity of air admitted was in most cases greater than the quantity required, and had a tendency to chill the gases, and the result was the air and smoke passed into the chimney without becoming ignited.

Another, and one of the latest attempts to burn smoke, is to introduce a portable or movable tube into every tube of the steam-boiler, for the purpose of separating the gas and smoke into thin sheets, in hopes the gas or smoke would become ignited. But the plan was impracticable, as it involved great trouble and expense.

Q. Do you believe that smoke can be successfully burned in common furnaces?

A. Not until the conditions under which it might be burned are more practically defined. First, the quantity of air that passes through the furnace more fully understood; second, the temperature of the furnace itself better known; third, the quantity of air required to burn different kinds of fuel,—three conditions that are very imperfectly understood by theorists, experts, and engineers at the present day.

GRATE-BARS.

PERFECT combustion is the starting-point in the generation of steam; the conversion of coal and air into heat must be the first process, and the second is to apply that heat with full effect to the boiler. The oxygen of the air is the only supporter of combustion, and the rate of combustion produced and the amount of heat generated in the furnace depend on the quantity of air supplied, and the quantity of air admitted depends on the size of the opening through which it passes. Then, as a matter of course, the grate-bars that offer the least obstruction to the air passing through them, and afford the largest area for the air combined with an equal distribution of the same, must be the most perfect for the purposes of combustion.

Grate-bars to be efficient must have a narrow surface exposed to the fire, and the spaces for admitting air must be numerous and well distributed to produce perfect combustion of the fuel.

IMPROVED GRATE-BARS AND BEARER.

The accompanying cut represents KEARNEY'S GRATE-BARS AND BEARER, which, it is claimed, are

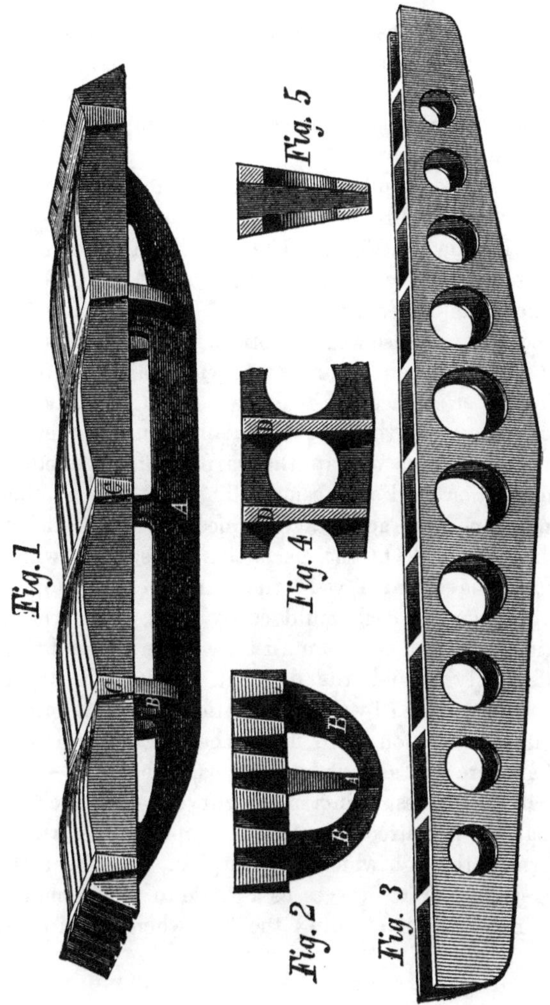
Fig. 1
A
B
C
Fig. 2
Fig. 3
Fig. 4
D
Fig. 5

not only of unusual durability, but also offer the advantage of a considerable saving in fuel. Fig. 1 shows a perspective view of the bars, of which a suitable number are joined together to form convenient-sized sections.

A is a longitudinal brace, to which are attached the transverse bridges, B B, of one of which an end view is shown in Fig. 2. The same illustration represents an end section of the bars, and the manner in which the latter are connected by the transverse blocks, C. It also will be noticed that the interstices or slots between the bars are widest at the bottom. The upper surface of the grate is corrugated, the object being to give an equal amount of metal at every point, and thus obviate the warping due to unequal contraction and expansion. There is also another and important advantage gained by this mode of construction. On the perfectly flat surface which would be afforded were the bars even on top, a thick layer of coal would easily pack, and, forming clinker, would make an air-tight covering, and thus effectually hinder the draught. This difficulty is entirely avoided by the corrugations, which admit of a free circulation of air under the fuel, from the fact that there will always be portions of the bars — generally the lowest points of the curves — on which the coal will not directly rest, so that open spaces will be formed, through which air can pass. Moreover, the irregular surface serves as a guide to the fireman to inform him, in cleaning the fire, when his slice-bar has reached the grate. The shape of the interstices between the bars, to which attention was directed above, is favorable to the ready passage of the ashes, while it aids in preventing clogging by clinker or otherwise.

The ends of the bars are open and bevelled as shown, the points of the extremities of two contiguous sections meeting on the upper surface of the bearer. This construction, as will be more clearly apprehended when considered in connection with the form of the bearer, by affording open ends, admits of a free circulation, and also prevents the bars from warping, and thus becoming useless before they are half worn out.

Fig. 3 represents a side view of a bearer on which the sections of grate rest. Figs. 4 and 5 are respectively longitudinal and vertical sections of the same. The bearer consists of two parallel bars pierced with a number of circular openings and connected together by transverse pieces, D D. The appliance is, therefore, in fact, a frame which, from the small amount of metal it contains, opposes but slight resistance to the passage of the draught. It is evident that a prominent merit of this invention is the ingenious combination of the hollow bearer and open ends of the sections of bars, so that the part of the grate which, in ordinary use, is the most liable to become packed and difficult to keep clean, is here as free and as clear as any other portion. A uniform circulation of air is consequently afforded through the entire length of the grate, and also a transverse current through the open supports on the under side.

This device has been thoroughly tested for some time, during which a continuous fire has been maintained. The result of two years' experiment at the Jersey City water-works, at Belleville, N. J., showed a direct saving of ten to fifteen per cent. in cost of both fuel and grates.

Any information concerning these bars can be had from WILLIAM KEARNEY, Belleville, N. J., or McFARLAND & McILRAVY, Newark, N. J.

DUTIES OF AN ENGINEER IN THE CARE AND MANAGEMENT OF THE STEAM-BOILER.

Q. What is the first duty of an engineer when he takes charge of an engine and boiler?

A. It is his duty to examine his boiler and see that the water is at the proper level.

Q. How much water should the boiler contain when in use?

A. The water should be kept up to the second gauge whilst working, and up to the third at night.

Q. Why should the level of the water be raised at night?

A. As a precaution against the water becoming too low from leakage or evaporation.

Q. In case the water should become dangerously low, what would be the duty of the engineer?

A. He should immediately draw the fire and allow the boiler to cool, and not admit any cold water to the boiler or attempt to raise the safety-valve, as it would be positively dangerous.

Q. Why would it be dangerous to raise the safety-valve?

A. Because it would lessen the pressure in allowing the steam to escape from the boiler, thus permitting the water to rise up and come in contact with the overheated iron, and probably cause an explosion.

Q. In case the water supply should be cut off from the boiler for a short time, what would be the duty of the engineer?

A. He should cover his fire with fresh fuel, stop his engine, and keep the regular quantity of water in the boiler until the accident is repaired and the water supply renewed.

Q. How should an engineer proceed to get up steam?

A. He should first see that the water is at the proper level; he should then remove all ashes and cinders from the furnace, and cover the grate with a thin layer of coal; and after placing his wood and shavings on the coal, he will be ready to start his fire.

Q. What advantage is it to place a covering of coal on the grate before the wood or shavings?

A. It is a saving of fuel, as the heat that would be transmitted to the bars is absorbed by the coal, and the bars are also protected from the extreme heat of the fresh fire.

Q. Should an engineer allow his fire to burn gradually when he commences to get up steam from cold water?

A. Yes; as by allowing the fuel to burn very rapidly, some parts of the boiler become expanded to their utmost limits, whilst other parts are nearly cold. Of course, a great deal depends upon the time in which he has to raise his steam.

Q. How should an engineer regulate his fire?

A. He should always keep the fire at a uniform thickness, and not allow any bare places or accumulations of ashes or dead coals in the corners of the furnace, as these places admit great quantities of cold air into the furnace and render the combustion very imperfect.

Q. Should an engineer avoid excessive firing as much as possible?

A. Yes; as excessive firing is always attended with more or less danger, because the intense heat repels the water from the surface of the iron and allows the boiler to be burned.

Q. How thick should an engineer keep his fires?

A. About 3 inches for anthracite coal and about 5 inches for soft coal; but he should regulate the thickness of the fire according to the capacity of the boiler; if the boiler is too small for the engine, the fire should be kept thin, the coal supplied in small quantities and distributed evenly over the grate, and the grate kept as free as possible from ashes and cinders; but if the boiler is extra large for the engine, the thickness of the fire makes but little difference.

Q. What should an engineer do in case, from neglect or any other cause, his fire should become very low?

A. He should neither poke nor disturb it, as that would have a tendency to put it entirely out, but he should place shavings, sawdust, wood, or greasy waste, on the bare places, with a thin covering of coal; then by opening the draught to its full extent, the fire will soon come up. If it should become necessary to burn wood on a cold fire, it is always best to make an opening through the coal to the grate-bars, so that the air from the bottom of the furnace can act directly on the wood and increase the combustion.

Q. Should an engineer give great attention to the regulation of the draught in the furnace?

A. Yes; the regulation of draught is one of the most important of an engineer's duties, for in fact it is next in importance to the regulation of the water in the boiler.

Q. How do you explain that?

A. Because it is well known that immense quantities of fuel are recklessly wasted by ignorance and carelessness in the management of the draught.

Q. How should an engineer regulate his draught to obtain the best results from the fuel?

A. He should have no more draught at any time than would produce a sufficient combustion of the fuel to keep the steam at the working pressure, as by opening the damper to its utmost limits great quantities of heat are carried into the chimney and lost.

Q. Can an engineer carry out this principle of regulating the draught in all cases?

A. No; only in furnaces and boilers that are sufficiently large to furnish the necessary amount of steam without forcing. Of course, where the boiler is too small for the engine, or has not sufficient heating surface, it is impossible to economize fuel.

Q. Do you consider it a good practice to throw a jet of steam under the furnace bars?

A. Only in some cases, where the draught is insufficient to produce the necessary combustion of the fuel; but it is considered an advantage, before cleaning the fire, to throw some water under the grate bars, as the oxygen from the steam thus generated under the furnace will unite with the oxygen of the atmosphere and insure a more rapid combustion of the fuel after the fire is cleaned.

Q. Is it objectionable to throw steam or water under the grate-bars of locomotive boilers, when such boilers are used for stationary engines?

A. Yes; as steam or water in the ash-pit forms a

lye with the ashes and corrodes the iron and destroys the water-legs of the boiler.

Q. Should an engineer in all cases keep his ash-pit clean?

A. Yes; by allowing the ash-pit to become filled with ashes and cinders the air becomes heated to a high temperature before entering the fire, which materially interferes with the combustion of the fuel; the grate-bars also become overheated, and in many cases either badly warped or melted down.

Q. How should an engineer keep his safety-valve?

A. He should keep it at all times in good working order, and move it at least once a day, particularly in the morning.

Q. Why should he move the safety-valve every morning?

A. To see that all its parts are in good working order before getting up steam.

Q. Would you consider it reprehensible conduct on the part of an engineer who would weight his safety-valve in order to carry a pressure greater than that he knew to be safe?

A. Yes; such conduct, if proved, ought to be sufficient to disqualify any engineer from ever taking charge of an engine and boiler again.

Q. What is the duty of an engineer in regard to blowing out his boilers?

A. He should carefully remove all the fire from the furnace, and see that the steam is at the proper pressure, say from 45 to 50 pounds. He should also close his damper.

Q. Should any time intervene between the drawing of the fire and the blowing out of the boiler?

A. Yes; at least one hour.

Q. Why should the blowing out of the boiler be deferred for an hour after the fire is drawn?

A. To allow the furnace to cool, and prevent the boiler from being injured with the heat after the water is all blown out.

Q. Why not blow out the boiler under a high pressure of steam, say 70, 80, or even 90 pounds to the square inch?

A. Because the higher the steam-pressure the higher the temperature of the iron, so that by blowing out the boiler under a high steam-pressure, the change is so sudden that it has a tendency to contract the iron and cause the boiler to leak.

Q. Should the engineer fill his boiler with cold water immediately after blowing out?

A. No; as the introduction of cold water into the boiler before the temperature of the iron becomes lower would in all probability cause the boiler to leak.

Q. How often should an engineer blow out his boiler?

A. Whenever he discovers any appearance of mud in the water.

Q. Is it not customary with some engineers and

owners of steam-boilers to blow out their boilers once a week?

A. Yes; but the wisdom of this practice is extremely doubtful, for, when fresh water is boiled, it is supposed to deposit its minerals, and after that it is not advisable to blow out the pure water and fill the boiler with water holding matter in solution and suspension; once in two or three weeks is as often as boilers ought to be blown out.

Q. Should an engineer, when filling his boilers, open some cock or valve in the steam-room of the boiler and allow the air to escape?

A. Yes; otherwise the air would retard the ingress of the water, and also collect in the steam-room of the boiler and prevent the regular expansion of the iron when the fire is started.

Q. What do you mean by the steam-room of a boiler?

A. I mean that portion of the boiler occupied by steam above the water.

Q. What is meant by the water-room in a steam-boiler?

A. That portion of the boiler occupied by water.

Q. What do you call the fire-line of the boiler?

A. The fire-line of the boiler is a longitudinal line above which the fire cannot rise on account of the masonry by which the boiler is surrounded.

Q. How often should an engineer clean the tubes or flues of his boiler?

A. At least once a week; he should also remove all ashes and soot that become attached to the outside of the boiler.

Q. What advantage is gained by cleaning the flues and tubes regularly, and also removing the soot and ashes that become attached to the boiler?

A. It makes a great saving in fuel, as it allows the fire to act directly upon the iron.

Q. How often should an engineer clean his boilers?

A. Every three months, if possible.

Q. Should an engineer, when cleaning his boilers, examine all stays, braces, seams, and angles of the boiler or boilers?

A. Yes; he should make a thorough examination of all parts of the boiler, seams, rivets, crown-sheet, crown-bars, crow-feet, cotters and braces; he should also sound the shell of the boiler with a very light steel hammer.

Q. Why should the engineer sound the boiler?

A. Because it is the only way in which he can determine the condition of the iron.

Q. How often should an engineer test his steam- or pressure-gauge?

A. At least once a year.

Q. Can an engineer test a steam-gauge himself?

A. No; unless he has a test-gauge, which is not very often the case. The gauge ought to be tested by another gauge built or made expressly for that purpose.

Q. How should an engineer keep his glass water-gauges?

A. He should keep them perfectly clean inside and out.

Q. How can an engineer clean his glass water-gauges inside?

A. By opening the drip-cock and closing the water-valve, and allowing the steam to rush down the glass and carry out the mud or sediment. They should also be swabbed out with a piece of cloth or waste on a small stick, when the boiler is cold; but care should be taken not to touch the inside of the glass with wire or iron, as an abrasion will immediately take place.

Q. In case an engineer has a glass water-gauge, should he neglect his gauge-cocks?

A. No; he should examine them several times in the day, see that they are in good working order, and grind or repair them if necessary. He should always be sure to shut them tight, as by leaving them loose the steam and water destroy the seat of the valve and render them useless.

Q. What evidence do dirty or broken glass gauges, filthy boiler-heads, leaking and muddy gauge-cocks give of a man's ability as an engineer?

A. They furnish strong evidence of his ignorance or neglect of duty.

Q. What should an engineer do in cold weather, when his pumps, boiler connections, steam-gauges, or water-pipes are liable to be frozen?

A. He should open all drip- or discharge-cocks and allow the water to run out when he stops work at night, and in the morning make a thorough examination of all steam and water connections before he starts his fires.

Q. In case it becomes necessary to stop the engine, and the steam commences to blow off at the safety-valve, what is the duty of the engineer?

A. He should immediately start his pump or injector, and also cover his fire with fresh coal, so that the circulation might be kept up by the feed-water, and the extreme heat of the fire absorbed by the fresh coal, instead of being communicated to the iron of the boiler; and he should not attempt, under any circumstances, to interfere with the free escape of the steam through the safety-valve.

Q. Whenever the fire-door of the furnace is open, should the damper be closed, if possible?

A. Yes; the door and the damper should never be open at the same time, unless it is absolutely necessary, as the cold air, that would otherwise have to pass through the fire and become rarefied, rushes in through the open door above the fire and impinges on the tube and crown-sheets, and has a tendency to contract the seams and cause them to leak.

Q. In case it should become necessary to examine the check-valve while steam is on the boiler, how should it be done?

A. The stop-cock between the check-valve and

boiler should be first closed before any attempt is made to unscrew or remove the check. Any neglect to close the stop-cock might result in a serious accident.

Q. How should an engineer proceed to make a joint on the man-hole or hand-holes of his boiler?

A. He should first carefully remove all gum or other material from the seat or flange where the joint is to be made, so that the gasket may have a smooth and solid bearing before he commences to tighten the nut.

Q. Do you know any other important duty an engineer should consider himself bound to perform?

A. Yes; he should daily make a thorough examination of all safety-valves, pumps, injectors, and all steam and water connections.

Q. What should be said of an engineer that would allow his boiler and engine to run on from bad to worse, expecting some day to have a general overhauling, instead of making repairs as they were needed?

A. He should be considered totally unfit for the position of an engineer.

Q. When can it be said that an engineer has done his duty?

A. When he shows by his work that he has cared for everything connected with his engine and boiler in the best possible manner.

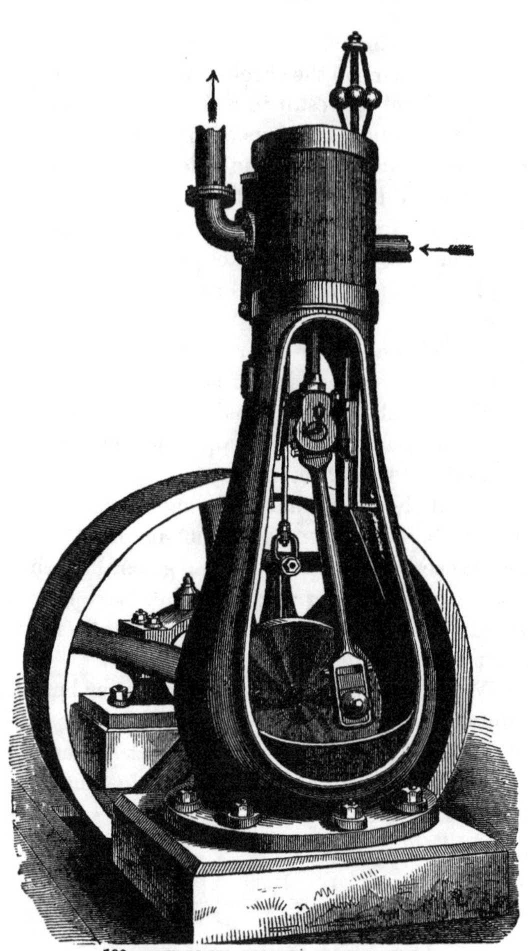

100 HORSE-POWER VERTICLE STEAM-ENGINE.
Kelly, Howell & Ludwig, Philadelphia, Pa.

HIGH-PRESSURE STEAM-ENGINES.

BEFORE the introduction of the steam-engine, horses were used to furnish power for various kinds of work, such as pumping water out of mines, raising coal, etc. When it was proposed to substitute the power of steam for that of horses, the proposal naturally took the form of furnishing a steam-engine capable of doing the work of a number of horses. Hence followed the usage of stating the number of horses which any particular engine was equal to.

Q. How is the power of a steam-engine generally expressed?

A. In horse-power.

Q. What is a nominal horse-power?

A. 33,000 pounds raised 1 foot high in 1 minute.

Q. Why is it that 33,000 pounds raised 1 foot high in 1 minute is adopted as a standard for a steam-engine?

A. Because, before the introduction of the steam-engine, it was found by experiment that with the average of horses the best speed for work was at the rate of 2½ miles per hour; and at that rate of speed a horse could raise perpendicularly a weight of 150 pounds 220 feet high in 1 minute, which is equivalent to 33,000 pounds raised 1 foot high in 1 minute, and was taken by Watt as a standard for horse-power, and is universally received as such. For instance; an engine of 60 horse-power can raise 33,000 pounds 1 foot high in a second, or an engine of 420 horse-power would raise 33,000 pounds 1 foot high in $\frac{1}{7}$ of a second.

Q. What is the indicated horse-power of an engine?

A. It is the power of an engine as shown by an instrument called an indicator.

Q. What is the actual horse-power of an engine?

A. It is the actual performance of the engine or the amount of power given out.

Q. What is the net horse-power of an engine?

A. It is the available power of an engine as determined by the indicator after deducting from it the power required to overcome the friction of the engine itself.

Q. What should be considered the unit of commercial horse-power in high-pressure steam-engines?

A. 4 circular inches area of piston, with an average pressure of 40 pounds to the square inch and a piston velocity of 400 feet per minute.

Q. What is the most convenient rule for finding the horse-power of an engine?

A. Multiply the area of the piston by the average pressure in pounds, less 5 pounds per square inch for friction; then multiply that product by the number of feet the piston travels per minute; then divide by 33,000 pounds. That will give the horse-power of the engine.

For example:

Diameter of cylinder in inches..........	10
	10
Square of diameter of cylinder...........	100
	.7854
Area of piston..................................	78.54
Pressure 60 lbs., cut-off ½ stroke......... Average press. 50 lbs., 5 off for friction	45
	39270
	31416
	3534.30
	250
	33000)883575.00
	26. horse-power.

Another Rule for finding the Horse-power of an Engine.

Multiply area of piston by boiler pressure, and this product by travel of piston in feet per minute; divide this last product by 33,000; then deduct 13 per cent. for loss by friction and condensation.

For example:

Diameter of cylinder in inches	12
	12
Square of diameter of cylinder	144
	.7854
Area of piston	113.0976
Pressure	60
	6785.856
Travel of piston	300
	33000)2035756.800
	61.689
Percentage off for friction, etc	.13
	8.01957
Horse-power of engine	53.669

Q. What do you mean by "average pressure"?

A. I mean, if the pressure is 60 pounds per square inch, and cut off in the cylinder at ½ stroke, the average pressure for the whole length of the stroke would be 50 pounds; if cut off at ¾ stroke, it would be 57 pounds.*

Q. How do you find the area of the piston?

A. Square the diameter of the cylinder, and multiply the product by the decimal .7854.

Q. Why do you square the diameter of the cylinder and multiply by .7854?

A. Because, when we square the diameter of the

* See Table of Averages, page 116.

cylinder, we get square inches, and when we multiply by .7854, we get the circular inches.

Q. What is the meaning of the word "area"?

A. It is the amount of surface exposed to the action of the steam.

Q. How do you find the travel of the piston in feet per minute?

A. We first find the length of the stroke of the engine, and the travel of the piston is double the stroke.

Q. What is the stroke of an engine?

A. Double the distance between the centre of the crank-pin and the centre of the crank-shaft. For instance, if it is 12 inches between the centre of the crank-pin and the centre of the crank-shaft, the engine is 2 feet stroke.

Q. Is the stroke and a revolution of an engine just the same?

A. Yes; whether we call it a stroke or a revolution, the travel of the piston is the same.

Q. What do you mean by the effective pressure on the piston?

A. I mean that the piston is under the action of the pressure of the steam from the boiler on one side, and the back-action caused by the pressure of the atmosphere on the other side. The difference between the two pressures is the effective pressure on the piston, and the power developed will depend upon the number of square inches acted upon, and the speed of the piston in feet per minute.

Q. What ought to be the minimum speed of the piston of any engine?

A. 240 feet a minute.

Q. Is the pressure in the boiler and the pressure in the cylinder nearly equal in all cases?

A. No; the pressure in the cylinder is in many cases 3·less than the pressure in the boiler; for that reason, in calculating the power of whole-stroke engines, or engines that do not work by expansion, not more than $\frac{2}{3}$ of the boiler pressure should be taken as the effective pressure in the cylinder.

Q. From what cause does the difference between the pressure in the boiler and the pressure in the cylinder arise?

A. It arises from different causes; first, from a malconstruction of the steam-pipe and steam-ports; second, from loss by radiation and condensation; third, by the action of the governor; and fourth, by the bad condition of the piston.

Q. What is the most economical steam pressure to use in the cylinder of a high-pressure engine?

A. From 80 to 90 pounds to the square inch.

Q. Why should 80 or 90 pounds to the square inch be more economical than lower pressure, say, 40 or 45 pounds to the square inch?

A. Because the loss is greater in low pressure than it is in high, owing to the pressure of the atmosphere; for instance, if we have a pressure of 45 pounds to the square inch on the piston, the loss by atmospheric pressure is 15 pounds to the square inch, which is about $\frac{1}{3}$ of the pressure on the piston, leaving only 30 pounds for useful effect and to overcome the friction of the engine; if we have a pressure of 90 pounds to the square inch, the loss is only 15 pounds to the square inch, or about $\frac{1}{6}$.

Q. Has the above calculation any reference to the pressure on the boiler as indicated by the gauge?

A. No; it refers only to the pressure on the piston.

Q. Does it take any more fuel to carry steam at a pressure of 90 pounds to the square inch than it does at 50 pounds to the square inch?

A. No; not quite so much.

Q. If that is the case, why not carry a pressure of 100 pounds to the square inch, or more?

A. Because, to carry such high pressures, it would be necessary to have boilers of smaller diameters and of better material; they should also be carefully protected from radiation by good non-conductors.

Q. Does a steam-engine that is too large for the work to be done, and running at a high speed, waste fuel?

A. Yes; as, for instance, an engine of 40 horse-power doing the work of a 20 horse-power engine, and running at a high speed, the steam would have to be throttled down, say, from 60 pounds boiler pressure to 25 pounds on the piston, which would be a loss of nearly $\frac{3}{5}$ in fuel.

Q. Is it a common error to get an engine too large for the power required?

A. Yes; as the steam necessary to drive a 30 horse-power high-pressure engine with *no load*, would give more than 10 horse-power in a small engine. The cylinder of any engine should be of sufficient size to give the full power required, leaving a reasonable margin for variation in pressure, and for recuperative power under sudden increase of load, *and no larger.*

Q. Are large engines doing the work easily, and working at a low pressure, economical?

A. Only when the number of revolutions is reduced in proportion to the work to be done.

Q. Is it difficult to fix the proper size of an engine, unless we know the power needed?

A. Yes; it is almost impossible to fix the size of an engine, unless we know the number of machines to be run, and the power required to run them or any of them: for instance, if we wish to run 3 machines, and it requires 10 horse-power for each machine, the engine ought to be 34 horse-power; if we wish to run 10 machines, requiring 10 horse-power each, the engine ought to be 115 horse-power.

Q. Is it necessary to know the pressure that will be used, and also the speed at which the engine will be run, before we can fix the power of an engine?

A. Yes; until we know the pressure to be carried, and the velocity at which the piston will travel, nothing very definite can be said as to the power of the engine.

Q. Which would be the most practicable way of increasing the power of an engine?

A. First, by increasing the pressure in the boiler; second, by increasing the speed of the engine; third, by putting on a new cylinder, which should not in any case be more than 2 inches in diameter larger than the old one.

Q. What ought to be the maximum piston speed of non-condensing engines?

A. Although there is a limit to the velocity of piston speed for all engines, that speed must be determined by the size and construction of the engine, as it would not be safe to subject any engine to a higher pressure than that for which it was built, nor to run it at a higher speed than its moving surfaces would bear. The velocity of piston speed must range between 240 and 700 feet per minute, according to the circumstances of the case.

Q. What are the four conditions that influence the economy of a steam-engine?

A. First, pressure; second, expansion; third, speed; fourth, size of cylinder.

Q. What are the three practical conditions that insure the greatest economy of steam?

A. First, the highest pressure; second, the greatest number of revolutions; third, shortest point of cut-off. If these three conditions are favorable, the highest point of economy is obtained.

Q. What is the common mode of applying the power of steam to the piston of the steam-engine?

A. One is to allow the steam to flow from the boiler to the cylinder during the whole length of the stroke; the other is to cut off the steam from the boiler when the piston has travelled a certain distance.

Q. What is the object of this last arrangement, viz., the closing of the communication between the cylinder and the boiler?

A. The great object of this arrangement is the saving of fuel.

Q. Does this account for some engines using more fuel and steam than others?

A. Yes; the reason why some engines use more steam than others of the same capacity is because the steam is not cut off or expanded.

Q. If the load on an engine will be such as to allow the steam to be cut off at a short point in the cylinder, what will be the effect?

A. The steam will be expanded to its full available limits, and a great saving will be made.

Q. If steam be applied to the piston the full length of the stroke, what will be the average pressure in the cylinder?

A. The average pressure will be as the pressure per square inch upon the piston.

Q. If steam, at 65 pounds to the square inch, be applied to the piston and cut-off at half-stroke, what will be the average pressure the whole length of the stroke?

A. 55 pounds to the square inch, being only 10 pounds less than the full pressure, or 16 per cent. of loss in powér, though half the quantity of steam was only used. This alone would effect a saving of 34 per cent. of fuel.

TABLE.

Showing the Average Pressure of the Steam upon the Piston throughout the Stroke, when cut off in the Cylinder from ¼ to ¾, commencing with 25 pounds and advancing in 5 pounds up to 100 pounds Pressure.

Steam Cut-off in the Cylinder.	*Pressure in pounds at the Commencement of the Stroke.*							
	25	30	35	40	45	50	55	60
	Average Pressure in pounds upon the Piston.							
¼	15	17¾	20¾	23¾	26¾	29¾	32¾	35¾
½	21	25¼	29½	33¾	38	42¼	46½	50¾
¾	24	28¾	33½	38½	43¼	48¼	53	57¾

Steam Cut-off in the Cylinder.	*Pressure in pounds at the Commencement of the Stroke.*							
	65	70	75	80	85	90	95	100
	Average Pressure in pounds upon the Piston.							
¼	38¾	41¾	44¾	47¾	50¾	53¾	56¾	59¾
½	55	59¼	63½	67¾	72	76¼	80½	84¾
¾	62½	67½	72¼	77¼	82	87	91¾	96½

Q. Which do you consider the most economical and available points at which to cut off steam in the cylinder?

A. ¼, ½, and ¾ of the stroke or travel of the piston. Beyond ¾, the saving is so small as to be hardly perceptible.

Q. What are the conditions under which steam can be worked expansively with the best results?

A. Every part of the steam-pipe, cylinder, and connections through which the steam flows from the boiler to the cylinder, must be carefully covered with some good non-conducting protectors ; for, unless the temperature of the steam and the temperature of the cylinder are kept nearly uniform, most of the benefits to be gained by expansion will be lost by radiation through the steam-pipe and the cylinder on the outside, and by condensation on the inside.

Q. Are the advantages arising from the working of steam expansively increased as we raise the pressure?

A. Yes ; steam at a pressure of 70 pounds to the square inch, worked expansively, will perform more than 7 times the duty of steam at 25 pounds per square inch.

Q. Is it necessary to make the cylinders larger in cases where steam is to be worked expansively ?

A. No ; with high pressures, by which expansion is most available, the cylinders of steam-engines can be smaller than they are usually made.

Q. When steam is worked expansively, which is the most effectual method of cutting off?

A. By the main valve.

LAP ON THE SLIDE-VALVE.

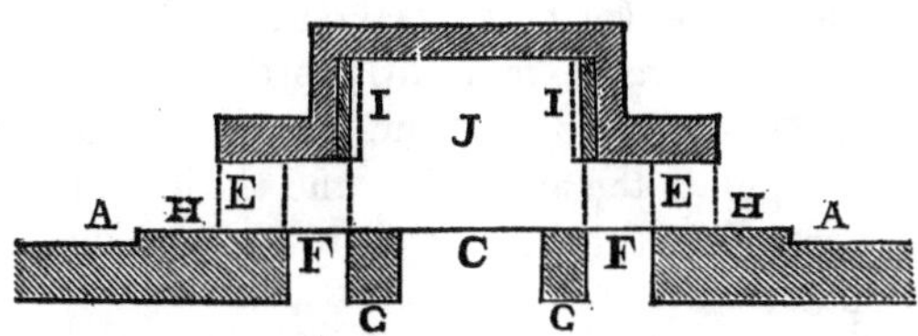

A A, Valve-seat. E E, Valve-face. F F, Steam-ports. G G, Bars. C, Exhaust-port. Dotted lines, H H, indicate outside lap. Dotted lines, I I, inside lap. J, Exhaust-cavity in valve.

Q. What is "Lap" on the slide-valve?

A. It is the lengthening of the valve to cut off the steam at ½ stroke, or any other point desired; for instance, when the valve is placed at ½ stroke over the port, the amount by which it overlaps each steam-port, either internally or externally, is known as lap. On the steam side it is named outside lap, on the exhaust side, inside lap; when the term lap and lead is used, it designates outside lap and lead.

Q. What is meant by inside clearance?

A. When the valve is so formed that at ½ stroke the faces of the valve do not close the steam-ports internally, the amount by which each face comes short of the inner edges of the ports is known as inside clearance.

Q. In what way is the distribution of the steam passing to and from the cylinder controlled?

A. By the outer and inner edges of the steam-ports and of the valve.

Q. Does the width of the exhaust-port, or the thickness of the bridges (or bars), make any difference?

A. No, if the valve is rightly proportioned; as the extreme edges of the steam-ports and those of the valve regulate the admission, and the inner edges of the ports and the valve control the exhaust, as before stated.

Q. How many distinct changes occur in the cylinder for every stroke of the piston?

A. Four: first, admission; second, suppression; third, release; and, fourth, compression.

Q. When does expansion of steam take place in the cylinder?

A. It takes place during the interval between the closing of the steam-port and the release of the exhaust.

Rule by which to ascertain the amount of lap necessary, on the steam side of a slide-valve, to cut the steam off at various fractional parts of the stroke.

To cut the steam off after the piston has passed through

$\frac{1}{2}$	$\frac{7}{12}$	$\frac{2}{3}$	$\frac{3}{4}$	$\frac{5}{6}$	$\frac{7}{8}$	$\frac{11}{12}$

of its stroke. Multiply the given stroke of the valve by

.354	.323	.289	.250	.204	.177	.144,

and the product is the lap of the valve in terms of the stroke.

A TABLE

Showing the amount of "Lap" required for Slide-valves when the Steam is to be worked expansively.

The traverse of the valves being ascertained, and also the amount of cut-off desired, the following Table shows the amount of "lap" required:

Traverse of the Valve in inches.	*The traverse of the piston where the steam is cut off.*							
	$\frac{1}{4}$	$\frac{1}{3}$	$\frac{5}{12}$	$\frac{1}{2}$	$\frac{7}{12}$	$\frac{2}{3}$	$\frac{3}{4}$	$\frac{10}{12}$
	The required "lap."							
2	$\frac{7}{8}$	$\frac{3}{4}$	$\frac{11}{16}$	$\frac{5}{8}$	$\frac{9}{16}$	$\frac{1}{2}$	$\frac{7}{16}$	$\frac{3}{8}$
$2\frac{1}{2}$	$1\frac{1}{16}$	1	$\frac{7}{8}$	$\frac{13}{16}$	$\frac{11}{16}$	$\frac{9}{16}$	$\frac{1}{2}$	$\frac{7}{16}$
3	$1\frac{1}{4}$	$1\frac{3}{16}$	$1\frac{1}{8}$	1	$\frac{15}{16}$	$\frac{3}{4}$	$\frac{5}{8}$	$\frac{9}{16}$
$3\frac{1}{2}$	$1\frac{1}{2}$	$1\frac{5}{16}$	$1\frac{3}{16}$	$1\frac{1}{8}$	$1\frac{1}{16}$	1	$\frac{7}{8}$	$\frac{3}{4}$
4	$1\frac{3}{4}$	$1\frac{9}{16}$	$1\frac{7}{16}$	$1\frac{5}{16}$	$1\frac{1}{4}$	$1\frac{1}{16}$	1	$\frac{13}{16}$
$4\frac{1}{2}$	2	$1\frac{13}{16}$	$1\frac{9}{16}$	$1\frac{1}{2}$	$1\frac{3}{8}$	$1\frac{1}{4}$	$1\frac{1}{8}$	$\frac{7}{8}$
5	$2\frac{1}{8}$	2	$1\frac{13}{16}$	$1\frac{9}{16}$	$1\frac{1}{2}$	$1\frac{3}{8}$	$1\frac{1}{4}$	1
$5\frac{1}{2}$	$2\frac{5}{16}$	$2\frac{3}{16}$	2	$1\frac{13}{16}$	$1\frac{5}{8}$	$1\frac{1}{2}$	$1\frac{3}{8}$	$1\frac{1}{8}$
6	$2\frac{1}{2}$	$2\frac{7}{16}$	$2\frac{3}{16}$	2	$1\frac{3}{16}$	$1\frac{5}{8}$	$1\frac{1}{2}$	$1\frac{3}{16}$
$6\frac{1}{2}$	$2\frac{3}{4}$	$2\frac{9}{16}$	$2\frac{7}{16}$	$2\frac{3}{32}$	2	$1\frac{13}{16}$	$1\frac{5}{8}$	$1\frac{1}{4}$
7	3	$2\frac{11}{16}$	$2\frac{9}{16}$	$2\frac{3}{8}$	$2\frac{3}{32}$	2	$1\frac{3}{4}$	$1\frac{7}{8}$
$7\frac{1}{2}$	$3\frac{3}{16}$	3	$2\frac{11}{16}$	$2\frac{1}{2}$	$2\frac{3}{8}$	$2\frac{3}{32}$	$1\frac{7}{8}$	$1\frac{1}{2}$
8	$3\frac{5}{16}$	$3\frac{3}{16}$	3	$2\frac{5}{8}$	$2\frac{1}{2}$	$2\frac{3}{8}$	2	$1\frac{5}{8}$
$8\frac{1}{2}$	$3\frac{5}{8}$	$3\frac{5}{16}$	$3\frac{3}{16}$	$2\frac{13}{16}$	$2\frac{11}{16}$	$2\frac{1}{2}$	$2\frac{1}{8}$	$1\frac{3}{4}$
9	$3\frac{13}{16}$	$3\frac{5}{8}$	$3\frac{5}{16}$	3	$2\frac{13}{16}$	$2\frac{11}{16}$	$2\frac{1}{4}$	$1\frac{7}{8}$
$9\frac{1}{2}$	4	$3\frac{13}{16}$	$3\frac{5}{8}$	$3\frac{3}{16}$	3	$2\frac{13}{16}$	$2\frac{3}{8}$	2
10	$4\frac{1}{4}$	4	$3\frac{13}{16}$	$3\frac{5}{16}$	$3\frac{3}{16}$	3	$2\frac{1}{2}$	$2\frac{1}{16}$
$10\frac{1}{2}$	$4\frac{7}{16}$	$4\frac{1}{4}$	4	$3\frac{1}{2}$	$3\frac{5}{16}$	$3\frac{1}{8}$	$2\frac{5}{8}$	$2\frac{3}{16}$
11	$4\frac{9}{16}$	$4\frac{7}{16}$	$4\frac{1}{4}$	$3\frac{5}{8}$	$3\frac{1}{2}$	$3\frac{3}{16}$	$2\frac{3}{4}$	$2\frac{1}{4}$
$11\frac{1}{2}$	$4\frac{13}{16}$	$4\frac{9}{16}$	$4\frac{7}{16}$	$3\frac{7}{8}$	$3\frac{5}{8}$	$3\frac{3}{8}$	$2\frac{7}{8}$	$2\frac{3}{8}$
12	5	$4\frac{13}{16}$	$4\frac{9}{16}$	$4\frac{1}{8}$	4	$3\frac{5}{8}$	3	$2\frac{1}{2}$

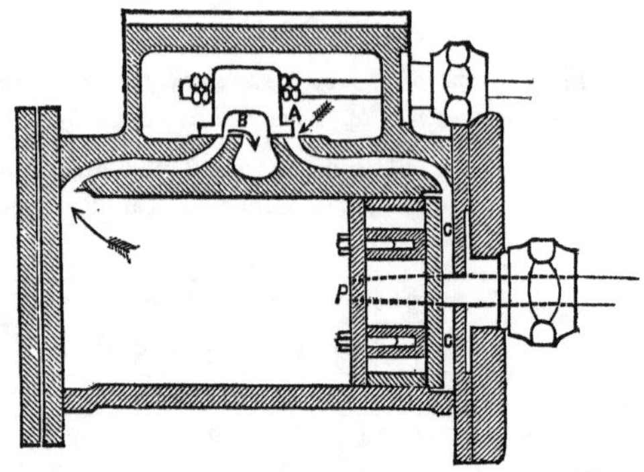

The arrow at A indicates outside lead. The arrow at B indicates inside lead. P is the piston at the beginning of the stroke. C C, clearance at end of the cylinder.

"LEAD" ON THE SLIDE-VALVE.

Q. What is "Lead" on the slide-valve?

A. It is the amount of opening the port has on the steam end when the engine is on the "dead centre."

Q. What do you consider the proper amount of "lead"?

A. The amount of "lead" on the valve must be determined by the circumstances of the case, as it is impossible to fix the exact amount for all engines; in most cases from $\frac{1}{32}$ to $\frac{1}{16}$ of an inch is sufficient, but in some engines it is necessary to give $\frac{1}{8}$ of an inch or even $\frac{1}{4}$ lead. The amount of lead for any engine will depend on the speed, the work to be done, the points of cut-off, etc.

Q. What is lead on the *exhaust?*

A. It is the amount of opening the exhaust-port has when the piston is at the end of the stroke, or, in other words, when the crank is on the dead centre.

Q. What do you consider the proper amount of lead on the exhaust?

A. In good practice, the amount of lead on the exhaust should be double the amount of lead on the steam-port. The exhaust in all cases ought to be liberated soon enough to preclude the possibility of back pressure, but exhaust-lead may be carried to excess. The proportioning of slide-valves presents so many complicated considerations, that it is impossible to give any definite instructions in any particular case without a full knowledge of the circumstances.

Q. What would be the effect on the engine if the exhaust is liberated too soon?

A. It would materially diminish the power of the engine; and if the engine was running at a slow speed under a heavy load, it might be difficult for the engine to complete the stroke, or, in other words, to pass its centres. The lead on the exhaust must be determined by the speed, the degree of expansion, and the amount of work to be done by the engine. For some classes of engines, such as pumping, propelling, or engines drawing heavy freight-trains up steep grades, steam must be forced into the cylinder to the last possible point of the stroke. Every engine should be specially arranged both for induction and eduction of steam, if the object is to get the full amount of power out of the engine.

Q. How can you determine accurately whether the exhaust opens at the right time or not?

A. Take off the "cap" or cover of the steam-chest, then disconnect the valve from the valve-rod; now take a short straight-edge and place it lengthwise on the edge of the exhaust-port; then with a sharp scribe lay off lines on the valve-seat, each side of the exhaust-port, that will appear above the valve; now lay the straight-edge on the valve-face and lay off similar lines on the exhaust-chamber that will appear on the edges of the valve; now place the valve on its seat, and give about $\frac{1}{32}$ of an inch lead; if the lines described on the valve-face are passed the corresponding lines on the valve-seat $\frac{1}{16}$ of an inch, the exhaust opens at the right time, if not, the exhaust-chamber in the valve ought to be cut away to give the necessary opening.

"CUSHION."

Q. What is "cushion"?

A. "Cushion" is steam admitted to or retained in the cylinder in front of the piston to overcome the *inertia* caused by the reciprocating parts of the engine. When the piston is cushioned with live steam it is done by giving an extra amount of lead. In most engines the exhaust steam remaining in the cylinder is sufficient to cushion the piston; but when it becomes necessary to cushion the piston with live steam, it should be done with caution, as any amount of cushion that retards the movement of the crank before it reaches the centre is a loss of power. Extreme cushion can always be discovered by a dull pounding noise in the engine; in such cases it is impossible to keep the packing tight around the piston-rod.

SETTING VALVES.

The proper and accurate *setting* of the valves of steam-engines is among the most important duties the engineer has to perform, as it involves a nicety of calculation and mechanical accuracy; no uniform rule can be adopted, as circumstances of construction, pressure, speed, points of cut-off, and work to be done, vary so much.

Q. How would you proceed to set the slide-valve of an engine?

A. Place the crank on the dead centre and give the valve the necessary amount of lead, then turn the engine on the other centre, and if the valve has the same amount of lead it is properly set. But if the lead on one end is more or less than on the other, the difference must be divided. When the valve is attached to the rod by means of jam-nuts great care must be taken not to jam the nuts against the valve, as that would prevent the valve from seating.

Q. In setting the slide-valve to cut off at any point of the stroke, is it necessary to move the eccentric?

A. Yes; the eccentric must be moved forward in proportion to the amount of lap added to the valve; whether steam is to be worked expansively or not, the valve must open at the same point of the stroke.

Q. What is the percentage of friction of the slide-valve as compared with the percentage of the cylinder's power?

A. It ranges from 10 to 30 per cent., according to the circumstances of the case; for while the cylinder's power decreases as the crank advances to the end of the stroke, the friction on the valve and the eccentrics increase until they absorb in some cases more than 25 per cent. of the power of the cylinder. The great aim of engineers for a long time has been to remove the weight caused by the pressure of steam from the back of the slide-valve, but it has been found almost impossible to produce a frictionless slide-valve.

Q. Is the slide-valve very imperfect and wasteful?

A. Yes; theoretically, I believe it only utilizes about 10 per cent. of the steam-power which ought to be produced from the combustion of good fuel; yet, imperfect as it is, on account of its simplicity of construction, durability and positive movement, it is the only valve now in use by all the railways in the world.

Q. What is the stroke of the slide-valve?

A. It is the distance that the valve moves on its seat.

Q. How do you find the stroke of the valve?

A. Move the valve in one direction to the extent of its stroke, and make a mark on the valve-rod; then reverse the movement to the opposite extremity and also make a mark; the distance between the two marks is the stroke or travel of the valve.

Q. Can the travel of the valve be increased or lessened by means of an intermediate bearing or slot in the arm of the *rocker?*

A. Yes; as the bearing in the rocker-arm that carries the eccentric hook is lengthened, we increase the stroke of the valve; if the same bearing is shortened, we lessen the stroke of the valve.

Q. How do you find the throw of the eccentric?

A. Measure the eccentric at the point which is called the throw or the heaviest side; now measure directly on the opposite or light side; the difference between the two measures will be the throw of the eccentric.

SIZE OF STEAM-PORT.

Q. What proportion should there be between the area of the steam-ports of a steam-engine and the area of the piston?

A. Steam-ports are made, in good practice, from

$\frac{1}{16}$ to $\frac{1}{10}$ the area of the piston; but they vary according to speed, etc.

SIZE OF STEAM-PIPE.

Q. What should be the diameter of the steam-pipe as compared with the diameter of the cylinder?

A. The diameter of the steam-pipe ought to be $\frac{1}{4}$ the diameter of the cylinder, but it varies on large engines. If the area of the steam-pipe, as compared with the area of the cylinder, is too small, it will cause *wire-drawing* of the steam and "priming" in the cylinder; for when the valve opens to admit steam to the cylinder, the rush of steam is so great that the water is carried over from the boiler to the cylinder, which is not only a great loss of power, but positively dangerous, as it occupies the clearance between the piston and the cylinder-head at the end of the stroke, and has a tendency to fracture the cylinder-heads.

SIZE OF EXHAUST-PIPE.

Q. What ought to be the diameter of the exhaust-pipe as compared with the diameter of the cylinder?

A. The diameter of the exhaust-pipe ought to be $\frac{1}{3}$ the diameter of the cylinder.

SIZE OF PISTON-ROD.

Q. What ought to be the diameter of the piston-rod as compared with the diameter of the cylinder?

A. The diameter of the piston-rod ought to be about $\frac{1}{6}$ the diameter of the cylinder, but when steel is used the diameter might be less.

MATERIAL FOR PARTS OF ENGINES.

Q. What kind of material do you consider the most suitable and durable to use in the manufacture of steam-engines?

A. For engines running at high speed, the piston-rod and crank-pin should invariably be made of steel; the boxes should be of brass in the proportion of 7 of copper to 1 of tin, and 1 of "spelter" to every 40 pounds of the mixture; the iron for cylinders and guides, if made from pig-iron, should be melted about 9 times, as that is the point at which cast-iron attains the most suitable density.

Q. What do you consider the most suitable material for the *gibbs* or *shoes* on the cross-head?

A. Wood; as hard brass or Babbitt metal has a tendency to wear the guides out of *true*, which incurs the expense of having them replaned on the face. Wood is free from the above objections, as it never wears or disfigures the guides; also it requires but little oil, and the cost is very trifling. *Lignum vitæ* and apple-wood answer a very good purpose for gibbs, the latter is preferable, as it is so much easier to work. Glass is sometimes used, and is said to answer very well for that purpose.

SPRING-PACKING.

Q. Does the setting out of the packing-rings in cylinders require great care and judgment?

A. Yes; for, like the setting of valves of steam-engines, no uniform rule can be laid down for that kind of work. If the packing is set out too tight, it diminishes the power of the engine, and by the increased friction has a tendency to flute or destroy both the surface of the packing and the inside of the cylinder. If the packing is too slack, and leaks, the steam occupies the cylinder in front of the piston, producing excessive cushioning, and materially diminishing the power of the engine.

PROPORTIONS OF ENGINES.

Q. What would you consider good proportions for steam-engines?

A. The diameter of cylinders of well-proportioned steam-engines ought to be about equal to the length of the crank between the centres; a 6-inch cylinder 12-inch stroke, 12-inch cylinder 24-inch stroke, 5-inch cylinder 10-inch stroke, 10-inch cylinder 20-inch stroke, as there is more vibration to long-stroke than to short-stroke engines. But the amount of work to be done and the speed of the piston must, to a certain extent, determine the diameter of the cylinder and the length of the stroke; engines intended to be run at high speed must of necessity be of short stroke.

Q. Are short-stroke engines more economical than long-stroke?

A. Yes; in all cases where steam pressure points of cut-off and travel of piston are the same.

REVERSING AN ENGINE.

Q. How would you proceed to reverse an engine?

A. First, make a mark on the side of the eccentric near the shaft, with a scribe or small chisel; make a corresponding mark on the shaft at the same point; then place one point of a pair of callipers on the mark on the shaft, and with the other point find the centre of the shaft on the opposite side; then with a scribe mark this point also; now unscrew the eccentric, and move it around in the direction in which the engine is intended to run until the mark on the eccentric comes into line with the second mark on the shaft; then make the eccentric fast, and the engine will run in the opposite direction.

Q. Does it make any difference in what position the crank is when the eccentric is moved?

A. No.

PUTTING AN ENGINE IN LINE.

Q. How would you proceed to put an engine in *line?*

A. First, remove the cylinder-heads, piston, cross-head and connecting-rod; now attach a small piece of iron or wood to the back flange of the cylinder;

fasten a small line to that finger and connect this line with a piece of iron or wood fastened in the floor at the back end of the bed-plate; now take a pair of inside callipers and find the distance between the line and the inside of the cylinder at four different points in the back and front ends of the cylinder; move the line attached to the stake at the back end of the bed-plate until all the points at the front and back ends of the cylinder are equally distant from the line; now move the crank up and see if the centre of the crank-pin feels the line; if so, move the crank on the other centre, and if the point at which the crank-pin strikes the line corresponds with the point on the other centre the engine is in line; if not, either the cylinder or the pillow-blocks will have to be moved, as the case may be.

Q. After the engine is "lined," how would you proceed to adjust the different parts?

A. First, insert the piston in the cylinder and slip on the packing-box and cross-head; now bring the centre of the piston-head to the centre of the cylinder by means of the set-screws or wedges, or whatever mechanical device may be used for that purpose; then screw up the follower-bolts; now lay a spirit-level on the piston-rod between the packing-box and cross-head. If the end of the rod next the cross-head is high, it can be lowered by slacking up the set-screws in the jaws of the cross-head. If the end next the cylinder is high, it can be levelled by screwing down the set-screws in the jaws; the piston-rod and cross-head should be levelled at both ends of the guides. Next place the box and strap on the wrist of the cross-head and put the connecting-rod in position and insert the key and gibb. Now move the crank to a convenient position and slip on the box and strap on the crank-pin and insert the key and gibb. It often happens in driving the keys on engines that they are adjusted too tightly, causing them to heat and destroy the bearings. The driving of keys on engines, like the adjusting of packing-rings in cylinders and the setting of valves, requires the exercise of great care.

SETTING UP ENGINES.

Q. How would you proceed to set up an engine?

A. Before commencing to set up an engine, it will be necessary to determine

First. The position or location the engine is to occupy in the shop or factory.

Second. The line of the main shafting in the building, if there be any; if not, the line of the building itself, at at least three different points in the direction in which the main shafting is to run; now line down from the centre of the main shaft, or from the line of the building, at two different points, to the floor on which the engine is to stand, and from these points line to the engine-shaft.

Third. Determine the height the bed-plate is to

stand above the floor; also the depth of the foundation, which in most cases would be about 3 feet below the level of the floor and 18 inches above.

Fourth. Make a *tamplet* the exact counterpart of the bed-plate, in which to hang the foundation-bolts; now set this upon four props at right angles to the main shaft in the building.

Fifth. Lay up the brick foundation to the level at which the engine is intended to stand; then remove the tamplet, and lower the bed-plate on to the foundation.

Sixth. Great care should be taken, when screwing down the foundation-bolts, to have the bed-plate perfectly level in every direction.

Seventh. A line should now be drawn exactly through the centre of the cylinder, and another line through the centre of the pillow-blocks; and if these two lines strike each other at right angles, the engine is properly set up.

Eighth. A straight-edge should now be placed across the bottom of the main bearings, in order to determine, by means of a spirit-level, whether the pillow-block boxes are perfectly level.

Ninth. The crank-shaft should now be placed in position and turned around several times, for the purpose of determining whether it bears equally in the boxes; if it does, the fly-wheel and driving-pulley can then be placed in position on their shafts, and all the other parts of the engine adjusted.

TABLE,

Containing the Circumferences and Areas of Circles from 4 to 26 inches in Diameter

Diam.	Circumfer.	Area.	Diam.	Circumfer.	Area.
4 in.	12.5664	12.5664	13/16	18.2605	26.5348
1/16	12.7627	12.9622	7/8	18.4569	27.1085
1/8	12.9591	13.3640	15/16	18.6532	27.6884
3/16	13.1554	13.7721	6 in.	18.8496	28.2744
1/4	13.3518	14.1862	1/16	19.0459	28.8665
5/16	13.5481	14.6066	1/8	19.2423	29.4647
3/8	13.7445	15.0331	3/16	19.4386	30.0798
7/16	13.9408	15.4657	1/4	19.6350	30.6796
1/2	14.1372	15.9043	5/16	19.8313	31.2964
9/16	14.3335	16.3492	3/8	20.0277	31.9192
5/8	14.5299	16.8001	7/16	20.2240	32.5481
11/16	14.7262	17.2573	1/2	20.4204	33.1831
3/4	14.9226	17.7205	9/16	20.6167	33.8244
13/16	15.1189	18.1900	5/8	20.8131	34.4717
7/8	15.3153	18.6655	11/16	21.0094	35.1252
15/16	15.5716	19.1472	3/4	21.2058	35.7847
5 in.	15.7080	19.6350	13/16	21.4021	36.4505
1/16	15.9043	20.1290	7/8	21.5985	37.1224
1/8	16.1007	20.6290	15/16	21.7948	37.8005
3/16	16.2970	21.1252	7 in.	21.9912	38.4846
1/4	16.4934	21.6475	1/16	22.1875	39.1749
5/16	16.6897	22.1661	1/8	22.3839	39.8713
3/8	16.8861	22.6907	3/16	22.5802	40.5469
7/16	17.0824	23.2215	1/4	22.7766	41.2825
1/2	17.2788	23.7583	5/16	22.9729	41.9974
9/16	17.4751	24.3014	3/8	23.1693	42.7184
5/8	17.6715	24.8505	7/16	23.3656	43.4455
11/16	17.8678	25.4058	1/2	23.5620	44.1787
3/4	18.0642	25.9672	9/16	23.7583	44.9181

TABLE.—(CONTINUED.)

Containing the Circumferences and Areas of Circles from 4 to 26 inches in Diameter.

Diam.	Circumfer.	Area.	Diam.	Circumfer.	Area.
5/8	23.9547	45.6636	7/16	29.6488	69.9528
11/16	24.1510	46.4153	1/2	29.8452	70.8823
3/4	24.3474	47.1730	9/16	30.0415	71.8181
13/16	24.5437	47.9370	5/8	30.2379	72.7599
7/8	24.7401	48.7070	11/16	30.4342	73.7079
15/16	24.9354	49.4833	3/4	30.6306	74.6620
8 in.	25.1328	50.2656	13/16	30.8269	75.6223
1/16	25.3291	51.0541	7/8	31.0233	76.5887
1/8	25.5255	51.8486	15/16	31.2196	77.5613
3/16	25.7218	52.8994	10 in.	31.4160	78.5400
1/4	25.9182	53.4562	1/8	31.8087	80.5157
5/16	26.1145	54.2748	1/4	32.2014	82.5160
3/8	26.3109	55.0885	3/8	32.5941	84.5409
7/16	26.5072	55.9138	1/2	32.9868	86.5903
1/2	26.7036	56.7451	5/8	33.3795	88.6643
9/16	26.8999	57.5887	3/4	33.7722	90.7627
5/8	27.0963	58.4264	7/8	34.1649	92.8858
11/16	27.2926	59.7762	11 in.	34.5576	95.0334
3/4	27.4890	60.1321	1/8	34.9503	97.2053
13/16	27.6853	60.9943	1/4	35.3430	99.4121
7/8	27.8817	61.8625	3/8	35.7357	101.6234
15/16	28.0780	62.7369	1/2	36.1284	103.8691
9 in.	28.2744	63.6174	5/8	36.5211	106.1394
1/16	28.4707	64.5041	3/4	36.9138	108.4342
1/8	28.6671	65.3968	7/8	37.3065	110.7536
3/16	28.8634	66.2957	12 in.	37.6992	113.0976
1/4	29.0598	67.2007	1/8	38.0919	115.4660
5/16	29.2561	68.1120	1/4	38.4846	117.8590
3/8	29.4525	69.0293	3/8	38.8773	120.2766

TABLE.—(Continued.)

Containing the Circumferences and Areas of Circles from 4 to 26 inches in Diameter.

Diam.	Circumfer.	Area.	Diam.	Circumfer.	Area.
$\frac{1}{2}$	39.2700	122.7187	$\frac{1}{4}$	51.0510	207.3946
$\frac{5}{8}$	39.6627	125.1854	$\frac{3}{8}$	51.4437	210.5976
$\frac{3}{4}$	40.0554	127.6765	$\frac{1}{2}$	51.8364	213.8251
$\frac{7}{8}$	40.4481	130.1923	$\frac{5}{8}$	52.2291	217.0772
13 in.	40.8408	132.7326	$\frac{3}{4}$	52.6218	220.3537
$\frac{1}{8}$	41.2338	135.2974	$\frac{7}{8}$	53.0145	223.6549
$\frac{1}{4}$	41.6262	137.8867	17 in.	53.4072	226.9806
$\frac{3}{8}$	42.0189	140.5007	$\frac{1}{8}$	53.7999	230.3308
$\frac{1}{2}$	42.4116	143.1391	$\frac{1}{4}$	54.1926	233.7055
$\frac{5}{8}$	42.8043	145.8021	$\frac{3}{8}$	54.5853	237.1049
$\frac{3}{4}$	43.1970	148.4896	$\frac{1}{2}$	54.9780	240.5287
$\frac{7}{8}$	43.5857	151.2017	$\frac{5}{8}$	55.3707	243.9771
14 in.	43.9824	153.9384	$\frac{3}{4}$	55.7634	247.4500
$\frac{1}{8}$	44.3751	156.6995	$\frac{7}{8}$	56.1561	250.9475
$\frac{1}{4}$	44.7676	159.4852	18 in.	56.5488	254.4696
$\frac{3}{8}$	45.1605	162.2956	$\frac{1}{8}$	56.9415	258.0161
$\frac{1}{2}$	45.5532	165.1303	$\frac{1}{4}$	57.8342	261.5872
$\frac{5}{8}$	45.9459	167.9896	$\frac{3}{8}$	57.7269	265.1829
$\frac{3}{4}$	46.3386	170.8735	$\frac{1}{2}$	58.1196	268.8031
$\frac{7}{8}$	46.7313	173.7820	$\frac{5}{8}$	58.5123	272.4479
15 in.	47.1240	176.7150	$\frac{3}{4}$	58.9056	276.1171
$\frac{1}{8}$	47.5167	179.6725	$\frac{7}{8}$	59.2977	279.8110
$\frac{1}{4}$	47.9094	182.6545	19 in.	59.6904	283.5294
$\frac{3}{8}$	48.3021	185.6612	$\frac{1}{8}$	60.0831	287.2723
$\frac{1}{2}$	48.6948	188.6923	$\frac{1}{4}$	60.4758	291.0397
$\frac{5}{8}$	49.0875	191.7480	$\frac{3}{8}$	60.8685	294.8312
$\frac{3}{4}$	49.4802	194.8282	$\frac{1}{2}$	61.2612	298.6483
$\frac{7}{8}$	49.8729	197.9330	$\frac{5}{8}$	61.6539	302.4894
16 in.	50.2656	201.0624	$\frac{3}{4}$	62.0466	306.3550
$\frac{1}{8}$	50.6583	204.2162	$\frac{7}{8}$	62.4393	310.2452

TABLE.—(Continued.)

Containing the Circumferences and Areas of Circles from 4 to 26 inches in Diameter.

Diam.	Circumfer.	Area.	Diam.	Circumfer.	Area.
20 in.	62.8320	314.1600	23 in.	72.2568	415.4766
1/8	63.2247	318.0992	1/8	72.6495	420.0049
1/4	63.6174	322.0630	1/4	73.0422	424.5577
3/8	64.0101	326.0514	3/8	73.4349	429.1352
1/2	64.4028	330.0643	1/2	73.8276	433.7371
5/8	64.7955	334.1018	5/8	74.2203	438.3636
3/4	65.1882	338.1637	3/4	74.6130	443.0146
7/8	65.5809	342.2503	7/8	75.0057	447.6992
21 in.	65.7936	346.3614	24 in.	75.3984	452.3904
1/8	66.3663	350.4970	1/8	75.7911	457.1150
1/4	66.7590	354.6571	1/4	76.1838	461.8642
3/8	67.1517	358.8419	3/8	76.5765	466.6380
1/2	67.5444	363.0511	1/2	76.9692	471.4363
5/8	67.9371	367.2849	5/8	77.3619	476.2592
3/4	68.3298	371.5432	3/4	77.7546	481.1065
7/8	68.7225	375.8261	7/8	78.1473	485.9785
22 in.	69.1152	380.1336	25 in.	78.5400	490.8750
1/8	69.5079	384.4655	1/8	78.9327	495.7960
1/4	69.9006	388.8220	1/4	79.3254	500.7415
3/8	70.2933	393.2031	3/8	79.7181	505.7117
1/2	70.6860	397.6087	1/2	80.1108	510.7063
5/8	71.0787	402.0388	5/8	80.5035	515.7255
3/4	71.4714	406.4935	3/4	80.8962	520.7692
7/8	71.8641	410.9728	7/8	81.2889	525.8375

For the circumferences and areas of circles of smaller diameters, see Table for Safety-Valves, page.

To find the circumferences of *larger* circles, multiply the diameter by 3.1416. For areas, multiply the square of the diameter by .7854.

RULES FOR THE CARE AND MANAGEMENT OF THE STEAM-ENGINE.

First. When the engine is stopped at night, the drip-cocks should always be left open in order to allow the condensed water to escape from the cylinder. They should not be closed for some time after the engine is started in the morning.

Second. Before starting the engine in the morning the cylinder should be warmed by admitting some steam, and moving the engine back and forth with the starting-bar.

Third. The oil or tallow should never be admitted to the cylinder until some time after the engine is started and the drip-cocks in the cylinder closed, as the tallow would otherwise be carried out by the condensed water and lost.

Fourth. All parts of the steam-pipe and the cylinder should be covered with cloth, felt, or some other good non-conductor to prevent radiation and condensation.

Fifth. Before packing the piston- and valve-rods all the old packing should be carefully removed and replaced with new packing, which should be cut in suitable lengths, and the joints placed at opposite sides of the box. The stuffing-box should then be screwed up until the leakage around the rod is stopped, and no further, as any unnecessary tightening of the stuffing-box will greatly diminish the power of the engine and soon destroy the packing by the increased friction.

Sixth. Piston-rod packing should always be kept in a clean place, as any dust or grit that may become attached to it has a tendency to cut or flute the rod.

Seventh. The spring-packing in the cylinder should always be kept up to its proper place, because, if allowed to become loose, the leakage materially interferes with the power of the engine. Setting out packing-rings requires the exercise of great care, because, if set too tightly, the friction produced will not only have a tendency to cut the cylinder, but will also perceptibly lessen the power of the engine.

Eighth. The piston should be removed from the cylinder at least twice a year, and the joints formed by the rings on the flange of the head and the follower-plate carefully ground with emery and oil. If badly corroded, they should be faced up in a lathe and made perfectly steam-tight.

Ninth. In driving the keys on the cross-head and crank-pin, the face of the hammer should never be used, unless the end of the key is protected by a piece of sheet-brass or copper.

Tenth. Great care should be taken, when packing the spindle of a governor, not to screw the packing down too tightly, as that would interfere with the free movement of the governor. All the parts of the governor should be kept perfectly clean and free from the gum formed by the use of inferior qualities of lubricating oils.

Eleventh. No more oil should be used on an engine than is absolutely necessary, as it is not only a loss, but often detracts from the appearance of the engine, and greatly interferes with its free and easy movement, from the accumulation of gum and dirt on its working parts.

Twelfth. In case the crank-pin should heat — which is a common occurrence with engines having a narrow bearing on the pin, but more particularly with engines that are slightly out of line — remove the key and slacken the strap and box; then pour in some flour of sulphur with a liberal supply of oil; then adjust the key, and the trouble will generally disappear.

Thirteenth. If the pillow-blocks of an engine should heat badly, remove the cap and pour in a good supply of pulverized bath-brick and water while the engine is in motion; after doing this for some time, wash out with oil, and wipe the bearing clean with waste, and it will be found to give permanent relief.

Fourteenth. In case any of the bearings of an engine should heat through the accumulation of matter deposited from the oil used, or sand, grit, or whitewash being dropped into the bearings, use a strong solution of concentrated lye with oil when the engine is in motion.

Fifteenth. A steam-engine should show, by its working and general appearance, that all its parts are thoroughly cared for.

Vertical Steam-Engine. Manufactured by Jacob Naylor, at the People's Works, Phila., Pa.

DIFFERENT KINDS OF ENGINES.

Q. What is meant by high-pressure or non-condensing engines?

A. High-pressure engines are a class of engines in which steam is used at a high pressure and exhausted into the open air. This class of engines includes locomotives and a great variety of stationary engines.

Q. What is meant by low-pressure or condensing engines?

A. Low-pressure engines are a class of engines in which steam is used at a low pressure and exhausted into a condenser, where a vacuum is produced by an air-pump; the piston being under the pressure of steam from the boiler on one side and the vacuum on the other, the difference between the two is the effective pressure on the piston. All ocean steamers, large river-boat engines, and a great many powerful factory engines, belong to this class.

Q. Is the vacuum perfect in low-pressure engines?

A. No; there is always more or less back pressure, caused by wear of the working parts or imperfections in the construction of the machinery.

Q. Are low-pressure engines more economical than high-pressure?

A. Yes; low-pressure engines possess great advantages over high-pressure in point of economy; but the first cost of a low-pressure is nearly double that of a high-pressure of the same power.

Q. What is meant by rotative engines?

A. Rotative engines are a class of engines in which the energy of the steam produces a continuous rotation of a shaft through the medium of a crank and a reciprocating piston. This class of engines includes a great variety of designs, namely, horizontal, vertical, beam, inclined, etc.

Q. What is meant by rotary engines?

A. Rotary engines are a class of engines in which a continuous rotation of a shaft is caused by the action of steam on a piston rotating within a steam-tight casing.

Q. Is the rotary engine as old as the rotative?

A. Yes; one was designed by James Watt, and there is hardly a treatise on the steam-engine in existence which does not contain an allusion to rotary engines; but the writers all agree that nothing would be gained by the substitution of the rotary for the rotative engine, on account of the great difficulty in making the parts steam-tight without producing extra friction.

Q. What is meant by beam-engines?

A. Beam-engines are a class of engines in which the reciprocating motion of the piston is transmitted to the crank through the medium of a beam commonly called a walking-beam. The beam-engine was for a long time in great favor with engineers and steam users on account of its graceful and nicely-balanced movements. But within the past few years beam-engines have been gradually superseded by upright and horizontal engines.

Q. Does the beam-engine possess any advantage over the horizontal engine?

A. Yes; there is less loss of power by friction in large beam-engines than in large horizontal engines, as the piston hangs plumb in the centre of the cylinder, and has a tendency to assist the downward movement of the stroke rather than to produce friction, as in horizontal engines. The cylinders of beam-engines are less liable to wear out of round than those of horizontal engines. But the first cost of beam-engines is more than that of horizontal of the same power.

Q. Is the beam-engine as old as any of the former?

A. Yes; when steam-engines were first introduced, more beam-engines were constructed than any other kind.

Q. What is meant by inclined engines?

A. Inclined engines are engines set at an angle or an inclined plane for convenience, or to give direct motion to some machine or machinery without the intervention of belts or gearing.

Q. What is meant by horizontal engines?

A. Horizontal engines are a class of engines in which the cylinder, piston, and guides are set horizontally with the bed-plate.

Q. What is meant by oscillating engines?

A. Oscillating engines are a class of engines in which the cylinder swings or vibrates on trunnions, by which the motion of the piston is transmitted directly to the crank, without the intervention of connecting-rod or cross-head.

Q. What advantage do oscillating engines possess over other engines?

A. None at all, except in the economy of space. The trunnions of oscillating engines soon become leaky, and are difficult to repair. The speed of oscillating engines is very limited. As a factory engine, oscillating engines might be said to be a failure.

Q. What is meant by poppet-valve engines?

A. Poppet-valve engines are a class of engines used almost exclusively in this country. They have four valves — two steam and two exhaust — which are moved by cams and levers. The poppet-valve engine embodies some very good principles, and when in good order works very economically. But a limit of speed is soon reached in the poppet-valve engine, as at a very high speed poppet-valves may not seat themselves promptly. Poppet-valve engines take their steam very rapidly, and probably have less back pressure than slide-valve engines.

Q. What is meant by slide-valve engines?

A. Slide-valve engines are a class of engines in which the steam is admitted to the cylinder by means of a slide-valve worked by an eccentric or cam.

Q. Do slide-valve engines possess any advantage over other engines in point of economy?

A. No; the slide-valve is very imperfect and

wasteful, as under ordinary circumstances it utilizes only about 4 per cent. of the steam-power which ought to be produced from the combustion of the fuel used, and not more than 10 per cent. under the most favorable circumstances. Yet, on account of its simplicity of construction, durability, and positive movement, and the fact that there is hardly any limit to its speed, the slide-valve engine has been enabled to compete with the best modern improvements in steam-engines?

Q. What is meant by Corliss engines?

A. Corliss engines are a class of high-pressure engines in very general use. Their design and construction embody the right principle for the steam-engine. They have four valves — two steam and two exhaust; they take steam very rapidly, and probably have less back pressure on the exhaust than any other class of engines in use. They work very economically when in good order; but the great drawback to the Corliss engine is that the valve-gear is complicated and expensive, liable to wear very rapidly, and is difficult to repair.

Q. What is meant by compound engines?

A. Compound engines are a class of engines which embody both the high- and the low-pressure principles. Steam is admitted into one cylinder and used at high-pressure, and then exhausted into another of much larger area and worked on the low-pressure principle. Such engines are said to be very economical, but they are complicated, and the first cost of them is very great.

Q. What is meant by side-lever engines?

A. Side-lever engines are a class of engines (used principally for pumping water) in which the motion of the piston is transmitted to the pump by side-levers or beams.

Q. What is meant by duplex engines?

A. Duplex engines are a class of engines which, like the compound, use steam on the high- and low-pressure principles, and allow a very great limit of expansion. They are principally used for pumping water. They work very economically, but, like the compound, are complicated and expensive.

Q. What is meant by Cornish engines?

A. Cornish engines are a class of engines in which the motion from the piston is communicated to the pump by means of a beam. They are used exclusively for pumping water in cities and mines. Cornish engines might be said to belong to that class of engines called single-acting engines, as the steam acts on only one side of the piston. Steam being admitted to the cylinder, the piston is forced down to the end of the stroke, when the steam escapes to the condenser. The opposite movement of the piston is accomplished by the inertia or weight on the pump.

Q. Are Cornish engines more economical than other pumping-engines?

A. No; the only advantage they possess over

other pumping-engines is that they admit of a very large measure of expansion in one cylinder.

Q. What is meant by Bull engines?

A. Bull engines are a class of direct-acting pumping-engines. These engines take their name from their inventor — Bull. They do not sustain a high reputation either for durability or economy.

Q. What is meant by trunk engines?

A. Trunk engines are a class of engines in which the cylinder is stationary, and the reciprocating motion of the piston is communicated directly to the crank without the intervention of connecting-rod or cross-head, through the medium of a trunk or hollow tube working through a stuffing-box. Trunk engines are very wasteful of steam, as the large mass of metal entering into the composition of the trunk, moving, as it does, alternately into the atmosphere and steam, must cool and condense a great part of the steam. The radiation from the interior of the trunk is also great.

Q. What is meant by caloric engines?

A. Caloric engines are a class of engines in which air is used as a motive power. The air is forced into the cylinder by means of an air-pump, and expanded by heat, by which means the piston is forced forward to the end of the stroke. Caloric engines are invariably single-acting engines, and their power is very limited.

KNOCKING IN ENGINES.

Q. What are the principal causes of knocking in engines?

A. Knocking in engines generally arises from the following causes:

First. Lost motion in the boxes on the cross-head, crank-pin, and pillow-blocks, and in the key of the piston-rod in the cross-head.

Q. What is the most effectual remedy for knocking arising from the above causes?

A. Take up lost motion by means of the key, or file off the edges of the boxes, if *brass-bound.*

Second. Knocking is sometimes caused by the crank being ahead of the steam, which in most cases can be relieved by moving the eccentric forward in order to give more lead on the valve.

Third. Knocking is caused in many cases by too much lead on the valve. The simplest remedy for this is to move the eccentric back so as to give less lead.

Fourth. Knocking is sometimes caused by the exhaust closing too soon. The best remedy for this would be to enlarge the exhaust-chamber in the valve.

Fifth. Knocking is in some cases produced by there not being sufficient clearance between the piston and the cylinder-head at the end of the stroke. The remedy for this kind of knocking would be to turn off the heads of the cylinder on the inside, so as to give more clearance.

Sixth. Knocking sometimes arises from the wrist of the cross-head and the crank-pin becoming worn out of *round.* The most effective remedy for this cause is to file the crank-pin and wrist perfectly round.

Seventh. Knocking is often induced by the cylinder not being *counter-bored* the necessary depth. In such cases the piston-rings wear a shoulder at each end of the cylinder, and whenever the keys are driven or the packing-rings set out, the edges strike these shoulders and cause the engine to knock. The most practical remedy for knocking arising from this cause is to *recounter-bore* the cylinder.

Eighth. Knocking is sometimes caused by the engine being out of line. The surest remedy for this kind of knocking would be to put the engine exactly in line.

Ninth. Knocking often arises from shoulders becoming worn on the ends of the guides in cases where the gibbs on the cross-head do not run over. The most reliable remedy for such knocking would be to replane the guides.

Tenth. Knocking is sometimes caused by the follower-plate being loose. The best preventive for such knocking is to bring the bolts up tight. To do so, it is sometimes necessary to remove the deposit of rust or grease in the bottom of the holes.

Eleventh. Knocking is very often caused by the packing around the piston-rod being too hard and tight. The most effectual remedy for that is to remove all the old packing from the box and replace it with new, and only screw the box up sufficiently tight to prevent the escape of steam; for any extra friction on the rod is a great loss of power, and has a tendency to destroy the packing.

Twelfth. The knocking heard in the steam-chest is sometimes caused by lost motion in the jam-nuts or yoke that forms the attachment between the valve and rod. The remedy for this would be to remove the cover of the steam-chest and readjust the jam-nuts on the valve-rod.

VACUUM.

Q. What is a vacuum?

A. The literal meaning of the term vacuum is space unoccupied by matter.

Q. Suppose the cylinder of a steam-engine be filled

with steam that is vaporized from a few drops of water, can it be said to be void of matter?

A. No; but condense that steam to its original bulk into water, and withdraw this water from the cylinder, and the space formerly occupied by the steam will be unoccupied; no matter remaining in the cylinder, then there is what is termed a vacuum.

Q. Suppose this operation to have taken place under the piston of a steam-engine, would there be any resistance to be overcome in the descent of the piston?

A. No; the pressure of the atmosphere alone, which is 15 pounds to the square inch, would suffice to force the piston down with a power equal to the degree of vacuum formed.

Q. Were the first steam-engines constructed on this principle?

A. Yes; the first steam-engines were constructed with the upper end of the cylinder open to the atmosphere; steam was then admitted below the piston to raise it; and this steam being condensed in the cylinder by the application of cold water, the pressure of the atmosphere alone caused the downward stroke of the piston.

Q. Suppose steam at 5 pounds pressure to the square inch above the atmosphere, or, in other words, 5 pounds pressure on the steam-gauge, (which in reality is 20 pounds to the square inch,) be applied to the piston of an engine under the conditions above stated, what would be the effect?

A. If the pressure of the atmosphere be 15 pounds to the square inch, there will be a pressure of 20 pounds exerted on the piston; but if the pressure of the atmosphere is only 14 pounds to the square inch, as is often the case, the pressure on the piston would be only 19 pounds to the square inch.

Q. How do you explain that?

A. Because the resistance of the atmosphere on the safety-valve would be less, and the steam in the boiler also less in proportion to the reduced pressure of the atmosphere.

Q. Does it often happen that low-pressure engines heavily loaded vary their speed with the varying pressure of the atmosphere?

A. Yes.

Q. Suppose that the vacuum is not perfect, (in practice it is never so,) and that there remains in the cylinder some uncondensed steam, the resistance of which is equal to 3 pounds to the square inch, what would be the effect on the engine?

A. Then the steam on the upper side of the piston, at 5 pounds to the square inch, above the pressure of the atmosphere, would act with an effective force of only 17 pounds to the square inch because the upper side of the piston having exerted upon it a pressure equal to a pressure of 20 pounds to the square inch and the under side of the piston had a resistance of 3 pounds to the square inch, the effective pressure would be only 17 pounds.

Q. Is a vacuum power?

A. No; all power in the steam-engine is derived from the pressure of the steam on the piston; if there is no resistance on one side of the piston, the whole pressure on the other side is available. Whenever there is resistance on one side of the piston, whatever the amount may be, it must be deducted from the pressure on the other side.

THE INDICATOR.

THE instrument known by the name of the Steam-Engine Indicator was invented by the celebrated James Watt. For a considerable period Watt kept the knowledge of that useful instrument to himself, but being obliged at last to send an engine abroad, and being responsible for its erection and proper working, he furnished a mechanic, whom he sent out to superintend the erection of the engine, with an Indicator, having previously instructed him in the use of the instrument.

Since the days of Watt, the Indicator has received several important improvements.

Q. What is the advantage of the Indicator?

A. The Indicator enables us to calculate with accuracy the pressure of steam exerted on the piston through the whole length of the stroke; it also shows at what part of the cylinder the piston is, and when the valves open or close.

Q. Is it possible to tell these things without the use of the Indicator?

A. No, not with any degree of accuracy. At a glance the Indicator reveals the inner working of the steam-engine, and by the use of the instrument, the good and bad qualities of the engine are registered on paper by the engine itself.

Q. Do you know of any other advantage to be gained by the use of the Indicator?

A. Yes; by means of the Indicator the owner of a factory may ascertain the whole power he is using, also the force required to overcome the friction of his engine and machinery, or the power required for any single room, or any particular machine, or any number of machines.

Q. Is there any other important knowledge we can obtain, concerning the steam-engine, by the use of the Indicator?

A. Yes; by means of the Indicator we are enabled to calculate how much more power an engine will exert by an increase of pressure, or by different degrees of expansion, according to the circumstances of the case.

Q. Would an extended knowledge of the use of the Indicator be of great importance to engineers and owners of steam-engines?

A. Yes; by a skilful use of the Indicator they could obtain a knowledge of the condition of every description of engines; also how to increase the power of each to its extreme limit of capability with the least expenditure of fuel.

Q. Why is it that the Indicator is not more extensively applied to steam-engines than it is at the present day?

A. First, because the Indicator, to a great extent, is in the hands of a class of men who try to deceive owners of steam-engines and steam-boilers by showing the enormous amount of work an engine is doing, for the purpose of magnifying the great evaporative powers of some patent steam-boiler. Second, the Indicator is also applied by some steam-engine builders (on engines of their own manufacture) who have an interest in showing that the engine is performing a far greater amount of work than that which was contracted for. Such men hardly ever work up their own cards, and even if they do, there is such a vast difference in the results of their trials on the same engines, that the owners of steam-engines are no better informed as to the condition of their engines, or the amount of work done by the engine, than they were before the Indicator was applied.

It is to be regretted that the benefit to be derived from the use of the Indicator is so imperfectly understood by engineers and owners of steam-boilers.

Bramwell, in his address before the British Association, claimed that one of the greatest incentives to economical working which owners of steam-engines could offer to engine-builders, would be the applica-

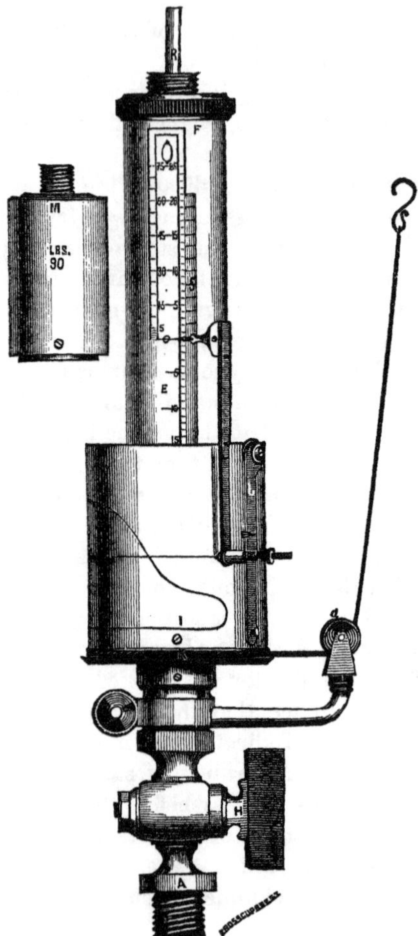
R
F
M
LBS.
90
E
I
H
A

tion of the Indicator; for he declared that the Indicator was to the proprietor of a steam-engine what the mariner's compass is to the captain of a ship navigating the sea.

Q. Will you explain how to apply the Indicator to steam-engines?

A. Yes; but in order to do so it will be first necessary to give a description of the instrument itself.

First. The tap, *H*, forms a communication between the cylinder of the Indicator and the cylinder of the engine. There is a small hole in the side of the tap which opens into the tap-plug, and when the tap is open to the cylinder of the steam-engine and the Indicator, this small hole is closed by the plug being turned with its perfect side against the hole. When the tap is closed, the connection with the engine-cylinder is cut off. The small hole in the tap is then open through the plug to the cylinder of the Indicator, so that any steam remaining between the piston of the Indicator and the plug of the tap may escape into the atmosphere, and allow the pencil attached to the piston of the Indicator to settle down to the atmospheric line.

Second. The cylinder of the Indicator is fitted with a piston, the rod of which is shown at *R*. This piston is accurately ground into the cylinder, thereby avoiding packing; and when properly oiled and cleaned, it is steam-tight, except at very high pressures, where perfect tightness is not required, as any small portion of steam which may escape cannot affect the bulk of the pressure in the cylinder. From this construction the piston works freely with little friction.

Third. The piston-rod, *R*, is attached to the spiral spring, *S*, within the tube or casing, *F*, placed above the steam-cylinder of the instrument. This spring is so adjusted that when the piston-index is forced one-tenth of an inch above the atmospheric line, 0, marked on the scale, *E*, it represents 1 pound pressure of steam. The pointer forced *upward* each tenth of an inch, up to 25 tenths, will represent as many pounds pressure to the square inch. In the same way, when the vacuum is formed in the engine-cylinder, the spring will be distended by the pressure of the atmosphere upon the upper side of the Indicator-piston, and the piston will be forced *downward* as many tenths of an inch as the degree of rarity, or the quantity of steam extracted from the engine-cylinder in pounds per square inch below the pressure of the common atmosphere.

Fourth. Affixed to the casing there is the scale, *E*, with the atmospheric line, 0, in the centre. The tenths below 0 to 15 indicate the vacuum, and above 0 to 25 tenths, the steam pressure above the atmospheric pressure.

Fifth. The pencil-holder, *G*, is attached to the piston-rod, *R*, through an aperture cut in the casing, *F*, to allow the pencil-holder to move up and down with the piston-rod of the Indicator. The pencil can be screwed backward or forward, to obtain the exact length required, and is adjusted by a spring to allow it to accommodate itself to any little inequalities there may be on the revolving cylinder or the paper. In all cases a soft and good pencil should be used. The less the pressure upon the paper, the less the resistance to the free action of the piston.

Sixth. *I* is the revolving cylinder outside the casing, *F*. This cylinder revolves on its own axis. The paper on which the diagram is to be taken is fixed around this cylinder, and held in its place by the clip, *J*. On the bottom of the cylinder there is a cord round the pulley, *K*, which, after passing over the swivel-pulley, *a*, is then attached to the cross-head of the engine. It will be obvious that the traverse of the cross-head will pull the cylinder as far round as the string thus travels. On the relaxation of the string, caused by the return movement of the engine-piston, the pulley will again take up the cord, because an internal spring, similar to the spring of a self-winding measuring-tape, is enclosed in the revolving cylinder. The traverse of the cross-head pulls the cylinder one way round and the spring the other. By means of the cord attached to the cross-head of the engine, there is thus produced a regular traversing motion of the cylinder, *I;* and as the pencil presses at the same time against the paper affixed to the cylinder, and also moves up and down with the Indicator-piston — which piston is propelled by the same force as the engine-piston — the piston will describe a diagram according to the circumstances.

Seventh. The area of the cylinder of the Indicator is a quarter of a square inch, and each tenth of an inch on the index represents 1 pound pressure to the square inch of the piston of the engine. The spring, *S*, compressed, shows the steam pressure. Distended, it shows the atmospheric pressure upon the piston of the Indicator caused by the formation of the vacuum in the engine-cylinder, varying in accordance with the pressure of the uncondensed steam left in the cylinder. This spring is so adjusted as to meet the requirements of the pressure as it increases. When the steam exceeds 25 pounds to the square inch above the pressure of the atmosphere, there is an additional spring enclosed in a case for higher pressures, up to 90 pounds to the square inch, or for a still further increased pressure, if required, which is to be screwed on to the top of the casing, *F*. This additional spring is represented at *M*. The piston-rod of the Indicator passes up the centre of the spring, *M*, when it is fixed on the top of the casing, *F*, and comes in contact with the top attached to the second spring; so that instead of the resistance of only one there is the resistance of two springs for high-pressure steam.

Eighth. On the scale, *E*, for high pressure, the distances are marked thirty to the inch for steam above 25 pounds — one-thirtieth of an inch representing 1 pound pressure to the square inch on the steam side. The scale on the vacuum side, ten to the inch, or one-tenth of an inch, represents 1 pound pressure to the square inch whether with high- or low-pressure steam. It should be borne in mind that whatever may be the pressure of the steam, high or low, when the engine is a non-condensing or high-pressure engine, the vacuum side of the Indicator is not required.

As to the Application.—The Indicator should be connected with the cylinder of the steam-engine by means of holes drilled and tapped into the clearance at each end of the cylinder, on the side opposite to the steam-ports; nipples should be screwed into these holes and elbows attached to their outer ends, in order to form a connection by means of a T, into which the Indicator should be screwed as near the centre of the cylinder as convenient, so that the pressure communicated from the steam-cylinder to the Indicator may be as uniform as possible.

The following diagram (No. 1) is taken from one of Wright's Cut-off High-pressure Steam-engines. It will be seen that it is very perfect. Attention is also called to the fulness of the expansion lines, *E*, above the theoretical curves, *T*; this it is claimed is due to the temperature of the steam in the cylinder being kept up beyond the point of cut-off by a steam-jacket. It is also claimed that the steam pressure in the cylinder from the commencement of the stroke to the cut-off point, *L*, is equal to the full boiler pressure.

DIAGRAM. No. 1.

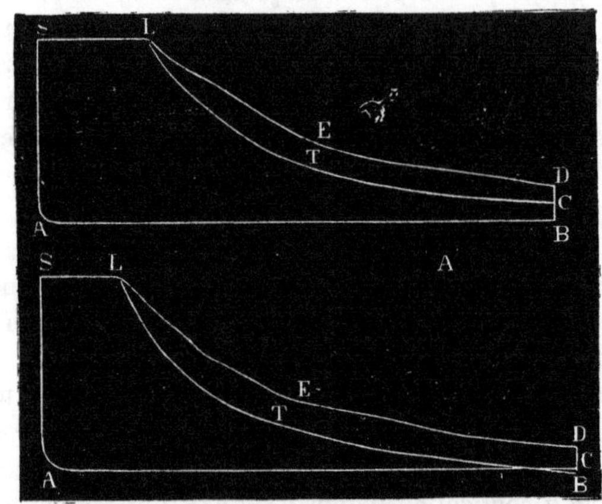

S L. Steam line. T C. Theoretical curve. E L. Expansion line.

DIAGRAM. No. 2.

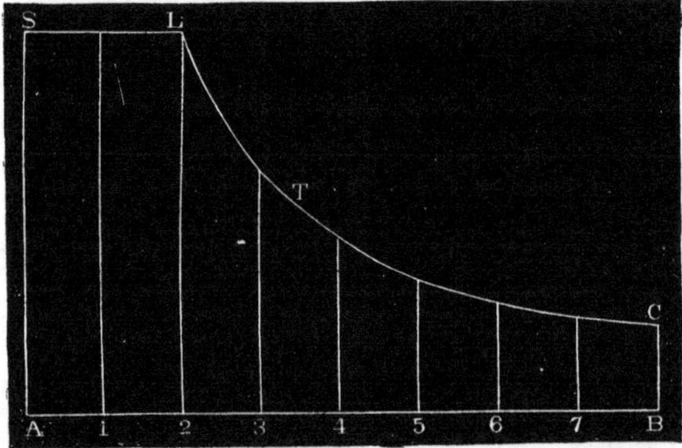

S L. Steam line. T C. Theoretical curve. 1, 2, 3, etc. Divisions of the Diagram.

A perfect Diagram. — According to Mariotte's law, the expansion-curve (No. 2, p. 163) should be a hyperbolic curve, where there are no extraneous circumstances to cause it to be otherwise; but unfortunately, in practice, this perfection is not attainable. Were Mariotte's law literally true—owing to the time required for the steam to enter and leave the cylinder, clearance of piston, leakage of valves and piston, and condensation in cylinder—it would be impossible to show a perfect diagram having all the corners well defined and the expansion line a true hyperbolic curve.

EXPLANATION.

A diagram with the steam-corner rounded shows the valve to have little or no *lead*, the steam being admitted upon the piston easily.

A diagram square and pointed shows the valve to have too much *lead;* that it opens too quickly, admitting the steam upon the piston with too great a force before the crank is in a position to receive it. This will in most cases cause the engine to tremble, and also the Indicator.

A diagram too much rounded at the commencement of the exhaust-corner shows want of *lead*, or a quicker opening of the valve on the exhaust side. In many instances the steam passages are too small; in such cases the exhaust ought to be open sooner than if the passages were of good proportion.

REMARKS.

It must be understood that every change in the engine which diminishes the area of the diagram by the rounding of its corners, diminishes the power of every stroke; for the space enclosed in the pencil-line exactly represents the power of a stroke of the engine, with the exception of the rounding of the *lead* corner.

While the expansion diminishes the power of each stroke, the reduced power of the engine is more than compensated for by the saving in the steam consumed.

RULE FOR COMPUTING THE POWER OF A DIAGRAM.

To compute the power of the diagram, set down the length of the spaces formed by the vertical lines from the base, in measurements of a scale accompanying the Indicator, and on which a tenth of an inch usually represents a pound of pressure; add up the total length of all the spaces, and divide by the number of spaces, which will give the mean length, or the mean pressure upon the piston in pounds per square inch; multiply the area of the piston in square inches by the pressure in pounds per square inch, and by the speed of the piston in feet per minute, and divide by 33,000, which gives the actual number of horse-power.

THE GOVERNOR.

In devices for regulating the speed of steam-engines we find that the principle of centrifugal force has received the most attention, and has been practically applied oftener than any other since the days of Watt, who first applied a Governor to the steam-engine.

The main object to be attained by the Governor is uniform speed of the engine under severe changes of machinery and varying steam pressure. A great amount of mechanical skill and capital have been expended, within the past few years, in efforts to improve the Governor, and the ingenuity of engineers and machinists has been taxed to make it sensitive without impairing its steadiness; and, as a result, there have been several Governors brought to notice, each of which is claimed to be a decided improvement on the Governor as first applied. The Governors constructed by Corliss, Porter, Judson, Shives, Jenkins, Conde, Pickering, and Brown are extensively used, and their merits are doubtless familiar to most engineers and steam users in this country.

CONDE'S AMERICAN GOVERNOR.

The cut on the opposite page represents Conde's American Safety-stop Governor. The novelties of this Governor consist in an improved double-acting valve, by which the Safety-stop is effected, and the

167

sensitive action given to the Governor. The compensating-wheel changes and regulates the speed of the engine as required, and the increased motion-movement regulates the velocity of the balls as they approach an outward plane. The valve, as shown in

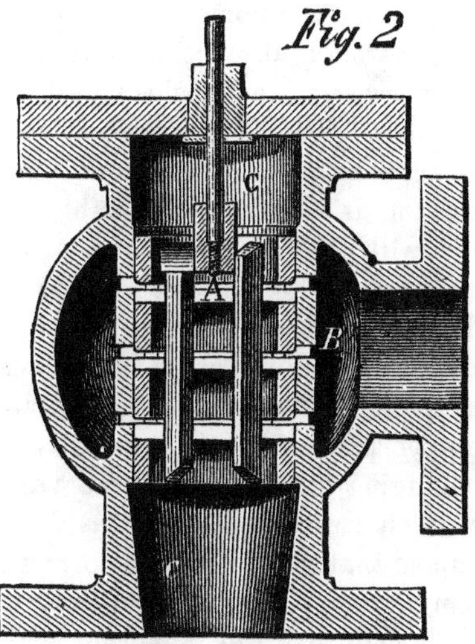

Fig. No. 2, consists of a series of rings (connected by internal ribs) forming three or more ports, through which the steam passes from similar ports formed in the chamber by a second series of rings connected by ribs within the globe. By thus constructing the valve with three or more ports, greater sensitiveness is gained; for in the downward movement of the valve three ports are cutting off steam instead of one, thus requiring only one-third the variation to do the same work, and but one-third the necessary change of engine speed to act on the Governor. The Safety-stop takes effect by the valve moving upward; the ports in the valve passing the ports in the chamber when the balls drop through any accident, from the belt breaking or otherwise, instantly closing all communication with the engine. This form of Safety-stop cannot fail to act so long as the Governor is in working order. This stop is simple and reliable, and in no way interferes with the free action of the Governor.

The compensating-wheel, as seen in the large cut, allows of a variation of speed in the engine at will. It is connected to the main frame of the Governor by a sleeve and screw, so that when operated either way, the whole upper part of the Governor is raised or lowered — changing the position of the valve in its seat, and letting in more or less steam, as the case may be. It is claimed that this is the only correct arrangement for changing the speed of an engine. But the most noticeable feature of this Governor is, that there are no levers, weights or springs to help its operation or complicate its parts; its operation is controlled by the simple and reliable principle of centrifugal force.

SHORT RULES FOR CALCULATING THE SIZE OF PULLEYS FOR GOVERNORS.

To find the diameter of Governor shaft-pulleys: Multiply number of revolutions of engine by diameter of engine shaft-pulley, and divide product by number of revolutions of Governor.

To find diameter of engine shaft-pulley: Multiply number of revolutions of Governor by diameter of Governor shaft-pulley, and divide product by number of revolutions of engine.

THE INJECTOR.

Of all the inventions of the mechanic and the scientist, nothing approximates so nearly to perpetual motion as the instrument now in general use and known as the Injector. It is one of the most beautiful productions of man's genius for the utilization of scientific purposes.

The Injector consists of a slender tube, through which steam from the boiler passes to another, or inner tube, concentric with the first. The latter tube conducts a current of water from a pipe into the body of the Injector. Opposite the mouth of this second tube, and detached from it, is a third fixed tube, open at the end facing the water supply-pipe, and leading from the Injector to the boiler.

The steam and water supply-pipes are fitted with

C
WATER
STEAM
GIFFARD'S INJECTOR
N°. 8. PAT. APR. 24. 1860.
B
TO BOILER

stop-valves, and the feed-pipe to the boiler with a check-valve.

When the instrument is ready for use, by simply opening the steam-valve steam enters the small steam-pipe and rushes out at its extremity, picking up the whole stream of water leaps across the open space with a loud hissing noise, and plunges with its burden of water into the open end of the feed-pipe at a tremendous velocity. Thus it will be seen that the steam that was admitted to the Injector from the boiler returns to the boiler, carrying with it more than twenty times its weight of water — not a drop of water is lost, not a particle of steam wasted.

Q. Can you explain the action of the Injector?

A. The principle on which the Injector works is that which is understood as the "lateral action of fluid." It was discovered by Venturi and Nicholson about 70 years ago. It is simply this: steam being admitted to the inner tube of the Injector, and the central conical valve being withdrawn, the steam escapes in a jet near the top of the inlet water-pipe. If the level of the water be below the Injector, the escaping jet of steam, by its superficial action (or friction) upon the air around it, forms a partial vacuum in the inlet-pipe. The water then rises in virtue of the external pressure of the atmosphere. Once risen to the jet, the water is acted upon by the steam in the same manner as the air had been seized and acted upon in first forming the partial vacuum into which the water rose.

Q. How can you account for the great surplus energy displayed in the working of the Injector, by which steam is taken from one boiler, at a pressure of, say 60 pounds to the square inch, and enabled to force water into another boiler under a pressure of 100 pounds to the square inch, or even more.

A. In this way: the velocity with which steam flows into the atmosphere is about 1500 feet per second; now let us suppose that steam is issuing with the full velocity due to the pressure in the boiler through a pipe an inch in area; the steam is condensed into water at the nozzle of the Injector without suffering any change in its velocity. From this cause its bulk will be reduced, say 1000 times, and, therefore, its area of cross-section — the velocity being constant — will experience a similar reduction. It will then be able to enter the boiler again by an orifice $\frac{1}{1000}$th part of that by which it escaped. Now it will be seen that the total force expended by the steam through the pipe, on the area of an inch, in expelling the steam-jet was concentrated upon the area $\frac{1}{1000}$th of an inch, and therefore was greatly superior to the opposing pressure exerted upon the diminished area.

TABLE OF CAPACITIES OF INJECTORS.

SIZE.	Size of Pipe for Connections.	PRESSURE OF STEAM IN POUNDS. 10	20	30	40	50	60	70	80	90	100
		CUBIC FEET OF WATER DISCHARGED PER HOUR.									
No. 2	½ in.	8.3	9.	9.7	10.4	11.1	11.8	12.5	13.2	13.9	14.6
" 3	¾ "	19.27	21.04	22.81	24.58	26.35	28.12	29.89	31.66	33.43	35.2
" 4	1 "	36.66	39.6	42.74	45.88	49.02	52.16	55.3	58.44	61.58	64.72
" 5	1¼ "	57.58	62.5	67.42	72.34	77.26	82.18	87.1	92.02	96.94	101.86
" 6	1¼ "	83.48	90.6	97.72	104.84	111.97	119.09	126.21	133.33	140.45	147.57
" 7	1½ "	114.03	123.75	133.48	143.2	152.93	162.65	172.38	182.1	191.83	201.55
" 8	2 "	149.2	162.	174.8	187.6	200.4	213.2	226.	238.8	251.6	264.4
" 9	2 "	189.2	205.35	221.51	237.66	253.82	269.97	286.13	302.28	318.44	334.59
" 10	2 "	233.84	253.8	273.76	293.72	313.68	333.64	353.61	372.57	393.53	413.49
" 12	2½ "	337.2	366.	394.8	423.6	452.4	481.2	510.	538.8	567.6	596.4
" 14	2½ "	451.49	491.45	531.41	571.36	611.32	651.27	691.23	731.18	771.14	811.09
" 16	3 "	600.32	651.6	702.88	784.16	805.44	856.72	908.	959.28	1010.56	1061.84

One nominal horse-power per hour will require 1 cubic foot of water per hour. When the pounds of coal consumed per hour can be ascertained, divide this by 7.5, and the quotient will be the quantity of water in cubic feet per hour.

Minimum capacity of Injectors about 50 per cent. of tabular capacity.

TEMPERATURES OF FEED-WATER.

Minimum temperature of feed admissible at different pressures of steam:

Pressure of steam (pounds) per square inch:

10	20	30	40	50	100
148°	138°	130°	124°	120°	110°

Q. Why is it that the temperature of the feed-water decreases as the steam pressure increases?

A. As the temperature of the steam increases with the pressure above a certain temperature of feed-water, it would be impossible to condense the steam in the Injector, and, as a consequence, the instrument would fail to work.

Q. Is the Injector more or less economical than the pump?

A. In point of economy, perhaps the advantage would be slightly in favor of the pump; but the advantages, if any, would be very trifling, which will be evident when we remember that with the Injector, as with the pump, waste heat may be employed to act upon the water between the feeding apparatus and the boiler, the difference being simply this: that with the pump the water will pass into the heating apparatus at its natural temperature, and will get all the heat it acquires before entering the boiler from the waste steam, while in the case of the Injector the water in passing through the instrument will acquire a certain amount of increase in temperature at the expense of the steam from the boiler before it comes in contact with the waste heat. This fraction, then, of the supply which might otherwise be derived from the exhaust steam, is all that is lost by the Injector. But it might be said, in favor of the Injector, that it requires no belt, packing, nor even repairs, and it occupies but very little space.

DOUBLE-ACTING BUCKET-PLUNGER STEAM-PUMP.

THE opposite cut represents an extremely simple and compact vertical steam-pump, provided with crank and fly-wheel. The pump is of the combined piston and plunger variety, and has but two valves; though possessing the same advantages as respects a steady delivery as the ordinary double-acting pump. It is intended for pumping all kinds of fluids. It is a double-acting piston-pump, the piston having a trunk on top. It has but one receiving and one discharge valve. The water is received only on the upward stroke, the amount being equal to the full capacity of the cylinder. Only one-half, however, is discharged, owing to the smaller area of the upper side of the piston. On the downward stroke, the water in the cylinder is forced out by the piston—one-half being discharged, the other half flowing into the upper end of the cylinder. These pumps are very superior as a fire-pump, for the reason that they will always start off at once when the steam is let on, which is very essential in a pump for this duty; as the time lost, in case of fire, to get an unreliable pump at work, is many times of more loss than the entire cost of a good pump.

They are made of any size to discharge from 5 to 5000 gallons per minute, and with either metal or rubber valves.

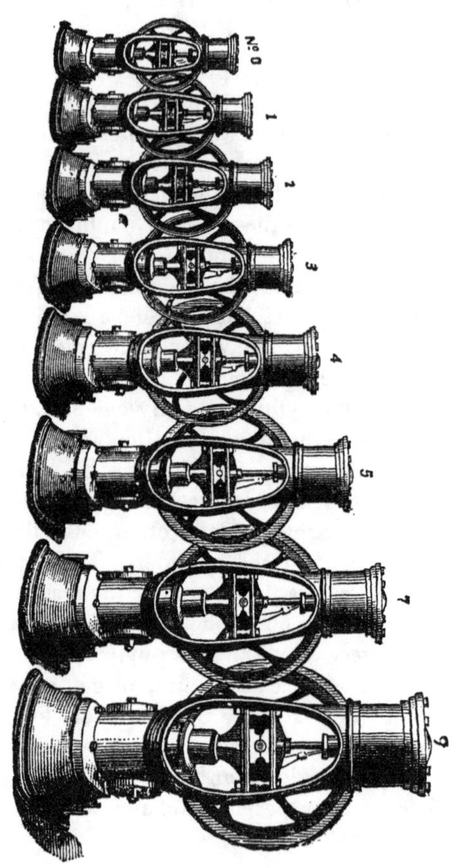

BUCKET-PLUNGER STEAM PUMPS,
VALLEY MACHINE COMPANY, EASTHAMPTON, MASS.

THE HEALD AND SISCO PATENT CENTRIFUGAL PUMPS—VERTICAL AND HORIZONTAL.

MANY devices have been employed for raising water and other liquids. Among the oldest and most general are those which depend, either wholly or in part, on atmospheric pressure. Of these are the pumps containing valves, and a tight piston moving in a straight cylinder. The numerous defects of this kind of pump long ago led the inventive mind to seek a more efficient substitute—a pump which, with a continuous flow, would raise more rapidly with less power, and would be more durable. The problem also required that *anything* capable of flowing should be raised as though it were clear water. Mud, sand, bark, etc., must prove no obstacle.

It was obvious, from the first, that the perfect pump must work without valves. The objections to valves are so manifest as to need no mention here. But without valves the pump could not have a tight piston. In short, the ideas of direct pressure and of atmospheric aid must be discarded, and a radically different system discovered, if possible.

Reflections of this sort, we should say, finally led to the building of pumps on the principle of the "Fan Bellows." This principle consists essentially in the rapid revolution of fans or arms in a scroll, sweeping or whirling the contained air to whatever vent might be found; the centrifugal momentum acquired in the revolution reacting from the inner walls of the scroll, and resolving itself into a force acting in the direction of the discharge.

In making a pump on this plan, no new principle was involved. But there was great merit in the novel application of the old idea. A long stride was made towards the solution of the problem when the first Centrifugal Pump was placed in the water.

The new pump was deservedly received with general favor. There were, however, certain defects still to be remedied — the principal one being a manifest waste of power and loss of efficiency, owing to the interference of the arms of the wheel with their own work. Reflection on this point led to the invention of the Hollow Arm Piston by Mr. Heald. This beautiful device was patented in 1865, and is already famous as one of the finest inventions of the age. The pump, of which it is the distinguishing feature, has taken first premiums at New Orleans (1871), Cincinnati and Brooklyn (1872), and also at the

Fair of the American Institute of the same year, after severe and protracted tests by the judges appointed for the purpose.

There are several thousands of these pumps in use by tanners, paper-makers, contractors, and wreckers, all over the United States, the Canadas, and Europe.

NOISELESS BOILER FEED-PUMP.

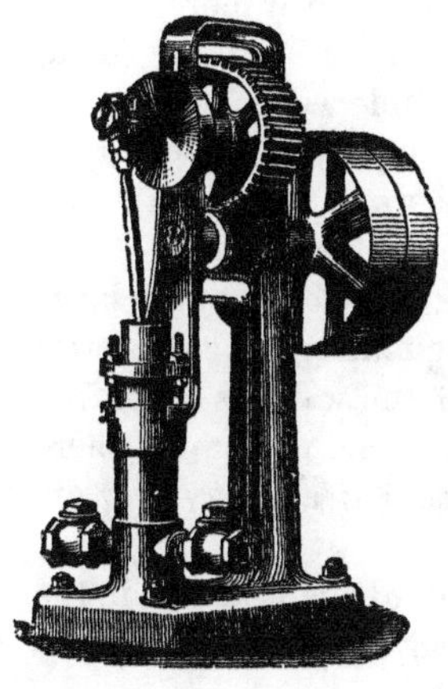

Diameter of Plunger.	Length of Stroke.	Strokes to the Gallon.	Diameter of Pulley.	Horse-power Boiler Feed.
$2\frac{1}{4}$ in.	4	12	12	20
$2\frac{1}{2}$	4	12	15	35
3	5	7	15	50
4	6	3	18	100
5	6	2	18	150
6	8	1	24	200

DIRECTIONS FOR SETTING UP STEAM-PUMPS.

Never use smaller pipes than the connections on the pump call for, and where long or crooked pipes are used, they should be larger. Use check-valve and strainer on the suction-pipe.

Run discharge-pipe of full size to the boiler, with check-valve between pump and boiler.

By reducing the diameter of a pipe ½, it diminishes its capacity to ¼.

A pipe 2 inches diameter, 100 feet long, will deliver but ¼ the quantity a pipe 2 inches diameter and 2 inches long, with same pressure.

Avoid angles, turns and bends in pipe where it is possible, as they retard the flow much more than is generally imagined. Where they must be used, make the turn on as large a circle as convenient.

If very hot water is to be pumped, the supply must be *above* the pump, so that the water will flow into it; as it is impracticable to raise it by suction.

Water expands by heating; therefore a pump large enough to furnish a given quantity of cold water would not be large enough if the water is heated.

Run the exhaust-pipe *down* from the pump where it is possible, that the condensed steam may flow off through it.

Care should be taken to guard against leaks in the suction-pipe, as a *very small* leak destroys the effectiveness of a pump.

Q. What should be the capacity of a pump or Injector to feed any particular steam-boiler?

A. As a cubic foot of water per hour is taken as a standard for a horse-power, the pump or Injector should be able to discharge more than a cubic foot of water per hour for each horse-power for which the boiler is rated; for instance, for a 10 horse-power boiler, the pump or Injector ought to be able to discharge 15 cubic feet of water per hour; for a 20 horse, 25 cubic feet of water, and so on in proportion.

A TABLE,

Containing the Diameters, Circumferences, and Areas of Circles, and the contents of each in gallons, at 1 foot in depth.

UTILITY OF THE TABLE.

EXAMPLES.

1. Required the circumference of a circle, the diameter being *five* inches.

In the column of circumferences, opposite the given diameter, stands 15.708 inches, the circumference required.

2. Required the capacity, in gallons, of a cylinder, the diameter being 4 feet and depth 10 feet.

In the fourth column from the given diameter stands 93.9754, being the contents of a cylinder 4 feet in diameter and 1 foot in depth, which being multiplied by 10 gives the required contents, 939¾ gallons.

3. Any of the areas in feet multiplied by .03704, the product equals the number of cubic yards at 1 foot in depth.

4. The area of a circle in inches multiplied by the length or thickness in inches, and by .263, the product equals the weight in pounds of cast iron.

Diam.	Circ. In.	Area. In.	Gallons.	Diam.	Circ. In.	Area. In.	Gallons.
1 in.	3.1416	.7854	.04084	6¼in.	19.635	30.679	1.59531
⅛	3.5343	.9940	.05169	⅜	20.027	31.919	1.65979
¼	3.9270	1.2271	.06380	½	20.420	33.183	1.72552
⅜	4.3197	1.4848	.07717	⅝	20.813	34.471	1.79249
½	4.7124	1.7671	.09188	¾	21.205	35.784	1.86077
⅝	5.1051	2.0739	.10784	⅞	21.598	37.122	1.93034
¾	5.4978	2.4052	.12506	7 in.	21.991	38.484	2.00117
⅞	5.8905	2.7611	.14357	⅛	22.383	39.871	2.07329
2 in.	6.2832	3.1416	.16333	¼	22.776	41.202	2.14666
⅛	6.6759	3.5465	.18439	⅜	23.169	42.718	2.22134
¼	7.0686	3.9760	.20675	½	23.562	44.178	2.29726
⅜	7.4613	4.4302	.23036	⅝	23.954	45.663	2.37448
½	7.8540	4.9087	.25522	¾	24.347	47.173	2.45299
⅝	8.2467	5.4119	.28142	⅞	24.740	49.707	2.53276
¾	8.6394	5.9395	.30883	8 in.	25.132	50.265	2.61378
⅞	9.0321	6.4918	.33753	⅛	25.515	51.848	2.69609
3 in.	9.4248	7.0686	.36754	¼	25.918	53.456	2.77971
⅛	9.8175	7.6699	.39879	⅜	26.310	55.088	2.86458
¼	10.210	8.2957	.43134	½	26.703	56.745	2.95074
⅜	10.602	8.9462	.46519	⅝	27.096	58.426	3.03815
½	10.995	9.6211	.50029	¾	27.489	60.132	3.12686
⅝	11.388	10.320	.53664	⅞	27.881	61.862	3.21682
¾	11.781	11.044	.57429	9 in.	28.274	63.617	3.30808
⅞	12.173	11.793	.61324	⅛	28.667	65.396	3.40059
4 in.	12.566	12.566	.65343	¼	29.059	67.200	3.49440
⅛	12.959	13.364	.69493	⅜	29.452	69.029	3.58951
¼	13.351	14.186	.73767	½	29.845	70.882	3.68586
⅜	13.744	15.033	.78172	⅝	30.237	72.759	3.78347
½	14.137	15.904	.82701	¾	30.630	74.662	3.88242
⅝	14.529	16.800	.87360	⅞	31.023	76.588	3.98258
¾	14.922	17.720	.92143	10in.	31.416	78.540	4.08408
⅞	15.315	18.665	.97058	⅛	31.808	80.515	4.18678
5 in.	15.708	19.635	1.02102	¼	32.201	82.516	4.29083
⅛	16.100	20.629	1.07271	⅜	32.594	84.540	4.39608
¼	16.493	21.647	1.12564	½	32.986	86.590	4.50268
⅜	16.886	22.690	1.17988	⅝	33.379	88.664	4.61052
½	17.278	23.758	1.23542	¾	33.772	90.762	4.71962
⅝	17.671	24.850	1.29220	⅞	34.164	92.885	4.82846
¾	18.064	25.967	1.35028	11in.	34.557	95.033	4.94172
⅞	18.457	27.108	1.40962	⅛	34.950	97.205	5.05466
6 in.	18.849	28.274	1.57025	¼	35.343	99.402	5.16890
⅛	19.242	29.464	1.53213	⅜	35.735	101.623	5.28439

Diam.	Circ. In.	Area. In.	Gallons.	Diam.	Circle.	Area.	Gallons.
11½in	36.128	103.869	5.40119	Ft. in.	Ft. in.	Feet.	1 ft. depth.
⅝	36.521	106.139	5.51223	2 5	7 7	4.5869	34.3027
¾	36.913	108.434	5.63857	2 6	7 10¼	4.9087	36.7092
⅞	37.306	110.753	5.75916	2 7	8 1⅜	5.2413	39.1964
				2 8	8 4½	5.5850	41.7668
Ft. in.	Ft. in.	In feet.	1 ft.depth.	2 9	8 7⅝	5.9395	44.4179
1	3 1⅝	.7854	5.8735	2 10	8 10¾	6.3049	47.1505
1 1	3 4⅝	.9217	6.8928	2 11	9 1⅞	6.6813	49.9654
1 2	3 8	1.0690	7.9944				
1 3	3 11	1.2271	9.1766	3	9 5	7.0686	52.8618
1 4	4 2⅛	1.3962	10.4413	3 1	9 8¼	7.4666	55.8382
1 5	4 5⅜	1.5761	11.7866	3 2	9 11⅜	7.8757	58.8976
1 6	4 8½	1.7671	13.2150	3 3	10 2½	8.2957	62.0386
1 7	4 11⅝	1.9689	14.7241	3 4	10 5⅝	8.7265	65.2602
1 8	5 2¾	2.1816	16.3148	3 5	10 8¾	9.1683	68.5193
1 9	5 5⅞	2.4052	17.9870	3 6	10 11⅞	9 6211	73.1504
1 10	5 9	2.6398	19.7414	3 7	11 3	10.0846	75.4166
1 11	6 2¼	2.8852	21.4830	3 8	11 6⅛	10.5591	78.9652
				3 9	11 9⅜	11.0446	82.5959
2	6 3⅜	3.1416	23.4940	3 10	12 5½	11.5409	86.3074
2 1	6 6½	3.4087	25.4916	3 11	12 3⅝	12.0481	90.1004
2 2	6 9⅝	3.6869	27.5720				
2 3	7 0¾	3.9760	29.7340	4	12 6¾	12.5664	93.9754
2 4	7 3⅞	4.2760	32.6976				

PISTON-ROD PACKING.

It is probable that on the whole, with steam-engines of plain construction, no part is more frequently out of order, and gives greater annoyance, than piston-rod packing. Hemp, when properly used, serves a good purpose, but its usefulness is limited, particularly where steam of a high pressure is used, as it soon loses its elasticity and, in consequence, becomes worthless. A vast deal of study and ingenuity have been applied to the removal of this annoyance and the production of a durable piston-rod packing. Wire-gauze, gum, soap-stone, jute, asbestos, and a great variety of other materials, have been tried, and with only partial success.

There has always been a general want of a permanent and reliable piston-rod packing.

Rule for finding the size of packing for piston or valve-rods:

Measure the piston or valve-rod; then measure stem of stuffing-box; divide the difference between them by two.

For example: rod 2 inches, box 4, packing 1 inch. Rod 1 inch, box 2, packing ½ inch. Rod ¾ inch, box 1½, packing ⅜. Rod 2 inches, box 3½, packing ¾. Rod 1½ inches, box 4 inches, packing 1¼.

INCRUSTATION.

All waters contain more or less mineral matter, which is acquired by percolation through the earth's surface, and consists principally of carbonate of lime and magnesia, sulphate of lime and chloride of sodium in solution, clay, sand, and vegetable matter in suspension.

Some waters contain far less mineral ingredients than others, such as rain-water, the water of lakes and large rivers, whilst wells, springs, and creeks hold large quantities in solution.

When such water is boiled, the carbonic acid is

driven off, and the carbonates, deprived of their solvents, are rapidly precipitated in a finely crystallized form, tenaciously adhering to the surface of the iron. Chloride of sodium, and all such soluble salts, are precipitated in the same way by supersaturation. This combined deposit, of which carbonate of lime forms the greater part, remains adherent to the inner surface of the boiler, undisturbed by the force of the most violent boiling currents. Gradually this accumulation becomes harder and thicker, until it is as dense as porcelain, thereby preventing the proper heating of the water by any fire that can be placed in the furnace. The high temperature necessary to heat water through thick scale will sometimes convert the scale into a substance resembling glass.

The evil effect of scale in steam-boilers is due to the fact that it is a non-conductor of heat. The conducting power of scale compared with that of iron is as 1 to 37; consequently, a greater amount of fuel is required to heat water in an incrusted boiler than if the same boiler were clean.

Scale $\frac{1}{16}$ of an inch thick will require an expenditure of fifteen per cent. more fuel. This expenditure increases as the scale becomes thicker; thus, when it is a quarter of an inch thick, sixty per cent. more fuel is needed to raise water in a boiler to any given heat. If the boiler is badly scaled, the fire surface of the boiler must be heated to a temperature according to the thickness of the scale.

For example: To raise steam to a pressure of 90 pounds, the water must be heated to a temperature of 324° Fah. If a quarter of an inch of scale intervenes between the shell and the water, it would be necessary to heat the fire surface of the boiler nearly 600°, or 100° Fah. above the maximum strength of iron. Now, it is a well-known fact, that the higher the temperature at which iron is kept, the more rapidly it oxidizes, and is made liable at any time to bulge or crack by internal pressure, and is often the cause of explosions.

Within the past few years a great many patent remedies for the removal and prevention of scale have been offered to the owners of steam-boilers, but, after a fair test of all these *remedies*, they have been nearly all abandoned, as not only useless, but even in many cases injurious, attacking and corroding the clean and sound iron, and producing no visible effect on the scale other than to change the color, and convey the impression that it was removed.

At a recent meeting of the American Railway Mechanics' Association, held at Louisville, Ky., the committee to whom was referred the subject of boiler incrustations, reported that they had prepared and issued, through the secretary of the association, a circular of questions to all the master mechanics of various railroads throughout the country, in order to elicit such information as they might possess on this subject. In compliance therewith, communications had been received from forty master mechanics; and although the number is small compared with the whole number of roads, yet it is nearly double that of any previous year, and the information so obtained is correspondingly extensive and valuable, confirming in substance the theory advanced in a paper read in convention last year, to the effect that the only effectual way to prevent incrustation is to purify the water, if possible, before it is allowed to enter the boiler. To this end the committee directed its efforts, and had given special attention to the reports of those who have experimented, with a view thereby of ascertaining the best and cheapest mode of accomplishing the same.

From all communications received, it is found that most of the roads located in the Eastern and Southern States are troubled but little with incrustation, while those in the Middle States are variously affected — some suffering greatly, others none at all. Western roads suffer most, many of them finding it necessary, in order to maintain average economy in fuel and reasonable safety to the boiler, to take out flues once in six to twelve months, for the purpose of removing scale from both boiler and tubes. Railway engineers in Western States realize similar difficulties in a greater or less degree, according to location. Mr. De Clerq, of the Toledo, Peoria, and Warsaw Railroad, reports having used batteries, and also many different kinds of powders, for the removal of incrustations, but all without any decided results. He finds distilled water the best of anything he has ever used to prevent lime incrustation. He thinks soft water, well filtered, will keep a boiler free from mud and scale.

Mr. Ham, of the New York Central, stated that he can run with economy, on the Eastern Division, four years without taking out the flues; while on the Middle Division, on account of lime and scale, he has to take them out, on an average, every year and a half, and on the Western Division every two years. He finds it necessary, on the Middle Division, to put new sheets in the bottom of the cylinder part of boiler on an average every five years; and, with good water, has only repaired that portion of the boiler once in eight to ten years. He has used batteries and powders, but finally abandoned them all. He is troubled with deposit of scale on the crown-sheet; gives crown-bars one inch clearance, and considers it as good as more space. He knows nothing equal to pure soft water to keep boilers free from mud and scale.

The following extract was taken from a report of a committee of the Railway Master Mechanics' Association, held in Boston in the summer of 1872. After a series of exhaustive experiments, the committee reported that the only preventive against incrustation was the use of pure water in steam-boilers. It was also stated that the introduction into the boilers of any of the so-called remedies now in use — whether they be batteries, powders, antilaminas, or filters — was comparatively useless for the removal or prevention of scale.

It was also stated that the extra expense in one year, from impure water and incrustation, would amount to $75,000 for every hundred locomotives. The committee considered that to boil sufficient water to supply a locomotive for one year, running 31,000 miles, would require an extra expenditure of $236 for fuel; but they considered that that was the only reliable means for preventing incrustation and all manner of ruptures and leaks in boilers.

Q. What is the most effective method for the prevention of incrustation?

A. The most effective method to prevent incrustation in boilers would be to boil the water in a tank, and then allow it to settle before using in the boiler. By this means the carbonates and minerals would become separated and deposited on the bottom of the tank, and the pure water could then be drawn off. Another effective method would be to blow out the boiler every evening, before the minerals that were held in suspension, by the agitation of the water, could settle and become attached to the sheets. A third would be to use rain-water, if possible. Any of the above three remedies would effectually prevent incrustation in steam-boilers, although it might be said that the two *former are* hardly practicable.

Q. What is the most effectual mechanical means that you know of for removing scale from boilers?

A. By picks and *scrapers.*

Q. Are boilers sometimes injured by this mode of cleaning?

A. Yes; the sheets of the boilers are sometimes cut by the use of the pick in the hands of ignorant or careless persons.

Q. Is this method practicable in all boilers?

A. No; only in cylinder boilers, as it is utterly impossible to remove scale from flue, tubular, or locomotive boilers by any mechanical means.

Q. What is the best preventive against scale becoming hard and firmly attached to the surface of the flues or tubes?

A. Never to let the surface of the flues or tubes become dry after the boiler is blown out. The boiler should always be again filled before the scale becomes dry.

Q. Mention some of the most common remedies resorted to for the purpose of preventing scale in boilers.

A. Indian meal, potatoes, oil-cake, molasses, gum caoutchouc, slippery elm, white-oak blocks, refuse logwood, and a variety of other vegetable substances, were at times tried without much success.

Q. What is the effect of vegetable substances on boilers?

A. They are sometimes as equally destructive to boilers as scale or incrustation.

Q. How do you explain that?

A. All vegetable matter contains more or less acid, and the iron is more open to attack from the acid than the scale is; so that some boilers suffer fearfully from the corrosion of the acid in the vegetable substances.

Q. What is the composition of most of the different patent boiler powders now presented to steam users for the prevention of scale?

A. Refuse logwood, sal soda, and yellow ochre.

Q. What is the effect of the above-named articles?

A. They have a tendency to deposit on the parts of the boiler most exposed to the fire; uniting the loose scales in solid masses, and causing the boiler to leak — very often to bulge or crack.

Q. Do you think it possible to separate the minerals from feed-water in the heater, by raising the water to a very high temperature, and forcing it through straw, shavings, or similar substances?

A. No; the capacity of any heater is so limited, and the time so short, that it is unreasonable to expect that any quantity of matter could be deposited. Even if a separation should take place in the heater, the matter will be carried into the boiler and be deposited on the iron.

Q. Is there any difference in scales formed under different pressures and different temperatures?

A. Yes; scales formed under low pressures and low temperatures are nearly always soft and porous; while scales formed under high pressures and high temperatures are hard and glassy.

Q. Would the removal of scale or incrustation be attended with economical results?

A. Yes; it would effect a saving of thousands of tons of coal annually, and a preservation of hundreds of lives destroyed by explosions that might be attributed to this cause alone.

Q. Do you think that pure water is the only effective means to prevent the accumulation of scale in steam-boilers?

A. Yes; for while several hundred patents encumber the records of the patent-offices of the United States and England, not one of them answers the purpose for which it was patented.

BOILER EXPLOSIONS.

THE risk of life and property involved in the use of the steam-boiler is still, as it has always been, a source of constant anxiety to the engineer and to the public.

Explosions continually take place, with the most disastrous results. Occurring without warning, and occupying but an instant of time, it is generally difficult, if not impossible, except in rare instances, to ascertain with certainty their true cause, as there is seldom a unanimous opinion on the part of experts who examine into the causes after the event.

But experience in the care and management of steam-boilers leaves no doubt in the minds of intelligent men but that the principal causes that operate to produce explosions are — inequality of expansion caused by some parts of the boiler becoming heated to a higher temperature than others; allowing the water to become dangerously low and pumping in cold water; sediment or incrustation being deposited on the parts of the boiler most exposed to the action of the fire, preventing the water from coming in contact with the iron, thereby allowing the plates to become overheated or burned; faulty construction; leakage, causing oxidation or rusting away of the iron; internal grooving; over-pressure; excessive firing; ignorance, recklessness, and mismanagement. The above causes include everything that experience and intelligence teach us to believe would cause a steam-boiler to explode; and it will be seen that the remedy for every one of these is within easy reach of the mechanic and the owner of steam-boilers. It might be said, without fear of contradiction, that the number of boilers that explode every year from all other causes are very few compared with those that occur from over-pressure and excessive firing.

It is also a well-known fact that a great many destructive steamboat and locomotive explosions have occurred just as the engine was starting, after standing still for some time. These, like all others, were the result of ignorance or carelessness. When an engine is stopped, and the communication between the cylinder and the boilers is closed, the whole heat of the fire acts on the plates of the boiler directly over and around the fire; the circulation ceases, and the iron and water become surcharged with heat to such an extent that, when the engine is started and the pressure on the surface of the water lessened, all the water on the surface of the iron directly over the fire immediately flashes into steam of tremendous elastic force. Whereas, if care had been taken to cover the fire with fresh coal, and the feed-pump or Injector had been started to force water into the boiler, the circulation of the water would have been kept up, and the heat that was transmitted to the iron would have been absorbed by the fresh coal and the feed-water.

The practical conclusion to be derived from all known facts in connection with the generation of steam is, that water in a boiler under some circumstances, — such as slow continued evaporation when a boiler is at rest,—can be nearly deprived of air, and the circulation being then feeble, portions of the water in contact with the plates may be heated to a higher temperature than that of the mass of water above it; under such circumstances, the sudden starting of the engine, or any other cause of agitation, producing an active circulation of the water, might cause a certain increase of the steam in such quantities, and of such elastic force, as would in many instances result in serious accident.

But that we shall continue to have frequent and

terrible steam-boiler explosions, there is no reason to doubt, so long as steam is used as a mere brute force, and so long as coroners' juries can be found to exonerate engineers and owners of steam-boilers from any blame, though it might be in evidence that the explosion, resulting in terrible loss of life, was brought about by their recklessness or avarice. The Lord's command, "Thou shalt not kill," limiting the power of masters over their servants, was one of the first steps towards the civilization of the world. But this solemn injunction does not seem to have much weight with some owners of steam-boilers in our days. Perhaps this arises from the fact that it is not in evidence that Moses was the owner of a steam-boiler.

Whenever a boiler does explode, which is a frequent occurrence in different parts of the country, it will be noticed that the self-styled experts are always on hand to ventilate their pet theories. Many of these theories are the merest vagaries, and only go to show the ignorance of their authors. The public read these speculations, and, in the absence of any intelligent refutation, swallow them down, and conclude that the whole subject is involved in dark mystery. Now, we claim that there is no mystery about steam-boiler explosions; they are all cause and effect. The City Inspection and the Hartford Inspection and Insurance Company have been demonstrating this fact for the past four or five years. Out of 3,000 steam-boilers in this city we did not have one explosion for over four years. Now, was it because engineers and owners of steam-boilers were more careful, and carried the water higher in their boilers than they had previously done, or was it because the City Inspector and the Insurance Company discovered defects that would lead to explosions, and compelled the owners of the boilers to repair them, and use their boilers at a safe working pressure?

It would be safe to say that the Hartford Steam-Boiler Inspection and Insurance Company have done more to prevent steam-boiler explosions in the last five years than had been done by all the laws passed by the National and State Legislatures for the past twenty years.

Few persons have any idea of the great internal strain continually exerted on the shells of boilers; for instance: in a boiler 54 inches in diameter, circumference 169 inches, pressure 85 pounds — a very common pressure to carry on steam-boilers — there would be continually exerted upon each inch in length of the shell a pressure of 14,365 pounds; and if the boiler is 18 feet, or 216 inches in length, the entire pressure on the shell of that boiler is 3,102,840 pounds or 1,556 tons.

In a boiler 36 inches diameter, circumference 113 inches, with a pressure of 70 pounds to the square inch, there would be a pressure of 7,910 pounds on every inch in length; and if the boiler was 14 feet in length, the whole pressure on the shell would be 664 tons.

STEAM- AND FIRE-REGULATOR.

THE numerous devices which have been employed by engineers for maintaining a uniform pressure of steam in boilers, shows the importance of a contrivance for this purpose. As a consequence, many steam- and fire-regulators have been introduced to the public, but most of them, from complexity or want of good workmanship, have failed to give satisfaction, and in many instances have proven themselves to be of more injury than advantage.

This Improved Regulator is self-adjusting, simple and durable in its construction, and not liable to derangement or loss of sensitiveness from time or use; having perfect control of the damper, it will, when once set to any required pressure, maintain that pressure in the boiler, without any attention from the engineer or fireman. In fact,

THE FOLLOWING ADVANTAGES ARE SECURED BY THIS REGULATOR:

First. Uniformity of pressure in the boiler within three pounds.

Second. Economy of fuel averaging ten per cent.

Third. Freedom from danger of explosion by excess of pressure.

The sizes of the Regulator are, No. 1, No. 2, No. 3, No. 4. The No. 2, with one ball at extreme end of lever, will close at 60 pounds per square inch; No. 3, in same position, 40 pounds per square inch. For higher pressures, extra balls are used.

CENTRAL AND MECHANICAL FORCES.

Q. What is centrifugal force?

A. It is the force with which a revolving body tends to fly from the centre.

Q. What is centripetal force?

A. It is the force that draws to the centre, or counteracts the centrifugal tendency.

Q. What is gravity?

A. Gravity is a downward pressure or weight.

Q. What is specific gravity?

A. It is the comparative density or weight that one body has to another of equal bulk.

Q. What is the centre of gravity?

A. It is that point in a body on which, if rested or suspended, the whole will remain in a state of equilibrium or rest.

Q. What is the force of gravity?

A. It is an accelerated velocity which heavy bodies acquire when falling from a state of rest.

Q. What is meant by the centre of oscillation?

A. It is a point in vibrating bodies in which all the force is collected.

Q. What is the centre of gyration?

A. It is the point in revolving bodies into which the momentum of the mass is concentrated.

Q. What is motion?

A. It is the effect of an impulsive force acting in such a manner as to impart linear or circular velocity by motive power.

Q. What is inertia?

A. It is that property of matter by which it tends when at rest to remain so, and when in motion to continue in motion.

Q. What is capillary attraction?

A. It is the property observable in all porous substances of raising water or other fluids above the natural level.

Q. What is friction?

A. Friction is the resistance experienced when one body is rubbed upon another, and is supposed to be the result of natural attraction.

Q. What are logarithms?

A. Artificial numbers which stand for natural numbers.

Q. How are the mechanical powers distinguished?

A. In the following order, viz.: lever, pulley, inclined plane, wheel and axle, and screw.

MENSURATION.

OF THE CIRCLE, CYLINDER, SPHERE, ETC.

1. The circle contains a greater area than any other plane figure bounded by an equal primeter or outline.

2. The areas of circles are to each other as the squares of their diameters.

3. The diameter of a circle being 1, its circumference equals 3.1416.

4. The diameter of a circle is equal to .31831 of its circumference.

5. The square of the diameter of a circle being 1, its area equals .7854.

6. The square root of the area of a circle, multiplied by 1.12837, equals its diameter.

7. The diameter of a circle multiplied by .8862, or the circumference multiplied by .2821, equals the side of a square of equal area.

8. The sum of the squares of half the chord and versed sine divided by the versed sine, the quotient equals the diameter of corresponding circle.

9. The chord of the whole arc of a circle taken from eight times the chord of half the arc, one-third of the remainder equals the length of the arc; or,

10. The number of degrees contained in the arc of a circle multiplied by the diameter of the circle and by .008727, the product equals the length of the arc in equal terms of unity.

11. The length of the arc of a sector of a circle, multiplied by its radius, equals twice the area of the sector.

12. The area of the segment of a circle equals the area of the sector, minus the area of a triangle whose vertex is the centre, and whose base equals the chord of the segment; or,

13. The area of a segment may be obtained by dividing the height of the segment by the diameter of the circle, and multiplying the corresponding tabular area by the square of the diameter.

14. The sum of the diameter of 2 concentric circles, multiplied by their difference and by .7854, equals the area of the ring or space contained between them.

15. The sum of the thickness and internal diameter of a cylindric ring, multiplied by the square of its thickness and by 2.4674, equals its solidity.

16. The circumference of a cylinder, multiplied by its length or height, equals its convex surface.

17. The area of the end of a cylinder, multiplied by its length, equals its solid contents.

18. The area of the internal diameter of a cylinder, multiplied by its depth, equals its cubical capacity.

19. The square of the diameter of a cylinder, multiplied by its length and divided by any other required length, the square root of the quotient equals the diameter of the other cylinder of equal contents or capacity.

20. The square of the diameter of a sphere, multiplied by 3.1416, equals its convex surface.

21. The cube of the diameter of a sphere, multiplied by .5236, equals its solid contents.

22. The height of any spherical segment or zone, multiplied by the diameter of the sphere of which it is a part, and by 3.1416, equals the area or convex surface of the segment; or,

23. The height of the segment, multiplied by the circumference of the sphere of which it is a part, equals the area.

24. The solidity of any spherical segment is equal to three times the square of the radius of its base, plus the square of its height, and multiplied by its height and by .5236.

25. The solidity of a spherical zone equals the sum of the squares of the radii of its two ends, and one-third the square of its height, multiplied by the height and by 1.5708.

26. The capacity of a cylinder 1 foot in diameter and 1 foot in length equals 5.875 of a United States gallon.

27. The capacity of a cylinder 1 inch in diameter and 1 foot in length equals .0408 of a United States gallon.

28. The capacity of a cylinder 1 inch in diameter and 1 inch in length equals .0034 of a United States gallon.

29. The capacity of a sphere 1 foot in diameter equals 3.9156 United States gallons.

30. The capacity of a sphere 1 inch in diameter equals .002165 of a United States gallon: hence,

31. The capacity of any other cylinder in United States gallons is obtained by multiplying the square of its diameter by its length, or the capacity of any other sphere by the cube of its diameter, and by the number of United States gallons contained as above in the unity of its measurement.

TO CALCULATE THE SPEED OF PULLEYS.

Example 1. To find the size of driving-pulley:

Multiply the diameter of the driven by the number of revolutions it should make, and divide the product by the revolutions of the driver. The quotient will be the size of the driver.

Example 2. The diameter and revolutions of driver being given, to find the diameter of the driven that shall make a given number of revolutions: Multiply the diameter of the driver by its number of revolutions, and divide the product by the number of revolutions of the driven. The quotient will be the size of the driven.

Example 3. To find the number of revolutions of the driven pulley: Multiply the diameter of driver by its number of revolutions, and divide by diameter of driven. The quotient will be the number of revolutions of the driven.

BELTING.

It is a common error among mechanics and owners of factories to make the face of their pulleys narrow, in order to economize on the first cost of belting; but this false economy seldom decreases the cost of the machinery, and only saves a trifle in the first cost of belting. The small amount saved is soon lost by the stopping of machinery caused by the slipping of belts, strain on the shafting, increased friction, requiring additional driving-power, and rapid destruction of the belts themselves. Were pulleys made of a proper size and width and face, and then covered with leather, and belts of proper width run with the *grain* side of leather to the pulley, thousands of tons of coal might be saved annually, and also an immense amount of trouble.

The importance of covering the face of pulleys with leather is realized by but few persons having charge of machinery; full 50 per cent. more work can be done without the belts slipping, if the face of the pulleys are covered with leather.

LEATHER BELTS.

Leather belts used with *grain* side to pulley will not only do more work, but last longer than if used with *flesh* side to the pulley; this is owing to the fact that the grain side is more compact and fixed than the flesh side, and more of its surface is brought in contact with the pulley. The smoother the two surfaces the less air will pass between the belt and the pulleys. The more uneven the surface of the belt and pulley the more strain necessary to prevent the belt slipping; for what is lost by want of contact must be made up by extra strain on the belt.

Leather belts, with grain side to pulley, can drive 34 per cent. more than flesh side.

A belt 1 inch wide, travelling 800 feet per minute over two smooth pulleys, will develop 1 horse-power.

A belt 1 foot wide, running at a speed of 70 feet per minute over smooth pulleys, will be equal to 1 horse-power.

A belt 3 inches wide, running at a speed of 280 feet per minute, will be equal to 1 horse-power.

A belt 5 inches wide, travelling 762 feet per minute over pulleys covered with leather, will develop 5 horse-power.

A belt 5 inches wide, in good condition, travelling over pulleys covered with leather, running at a speed of 1525 feet per minute, will transmit 10 horse-power.

LACING BELTS.

In lacing belts great care should be taken that the ends intended to butt together should be cut perfectly square; if not, the belt will stretch more on one side than the other, which will greatly impair its worth.

HORIZONTAL BELTS.

The driving half of horizontal belts should be the lower half when practical, as, when the belt stretches, the upper half will cover more of the pulley's surface. Long horizontal belts are better than short ones, as their weight increases their contact with the pulley.

PERPENDICULAR BELTS.

Belts running on pulleys perpendicular to each other should be kept tightly strained, as their weight tends to decrease their contact with the lower pulleys.

GREASING BELTS.

Belts, if dry or husky, should be greased with a mixture of neat's-foot oil and tallow, and dried in by the heat of the fire or sun.

RULES.

Rule for calculating the width of belts required for transmitting different numbers of horse-powers:

Multiply 36,000 by the number of horse-powers; divide the product by the number of feet the belt is to travel per minute; now divide the quotient by the number of feet, or parts of feet, of belt in contact with the smaller pulley; divide this last quotient by 6, and the result is the required width of belt in inches.

Rule for calculating the number of horse-powers a belt will transmit; its velocity, and the number of square inches in contact with the smaller pulley, being given:

Divide the number of square inches in contact with the pulley by 2; multiply this quotient by the velocity of the belt in feet per minute, and divide by 36,000; the quotient is the number of horse-powers the belt will transmit.

ACCIDENTS.

RULES FOR THE COURSE TO BE FOLLOWED BY THE BYSTANDERS, IN CASE OF INJURY BY MACHINERY, WHEN SURGICAL ASSISTANCE CANNOT AT ONCE BE OBTAINED.

If there is bleeding, do not try to stop it by binding up the wound. *The current of the blood to the part must be checked.* To do this, find the artery by its beating; lay a firm and even compress or pad

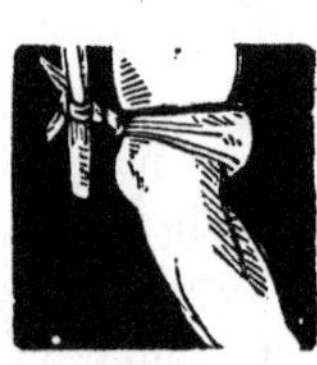

Fig. 1.

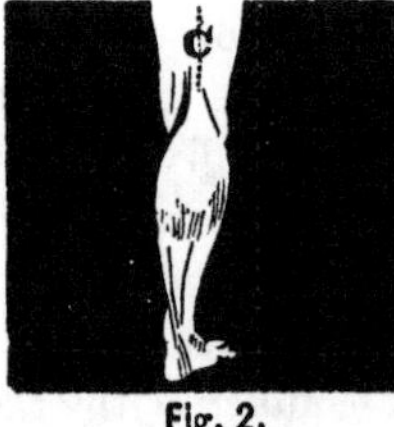

Fig. 2.

Fig. 3.

(made of cloth or rags rolled up, or a round stone or a piece of wood well wrapped) *over the artery, (see Figure* 1;) tie a handkerchief around the limb and compress; put a bit of stick through the handkerchief and twist the latter up till it is *just tight enough to stop the bleeding;* then put one end of the stick under the handkerchief to prevent untwisting, *as in Fig.* 3.

The artery in the thigh runs along the inner side of the muscle in front, near the bone. A little above the knee it passes to the back of the bone. In injuries at or above the knee, apply the compress high up on the inner side of the thigh, at the point where the two thumbs meet at C in *Figure* 4, with the knot on the outer side of the thigh. When the leg is inured below the knee, apply the compress at the back of the thigh just above the knee at C in *Figure* 2, and the knot in front, as in *Figures* 1 and 3.

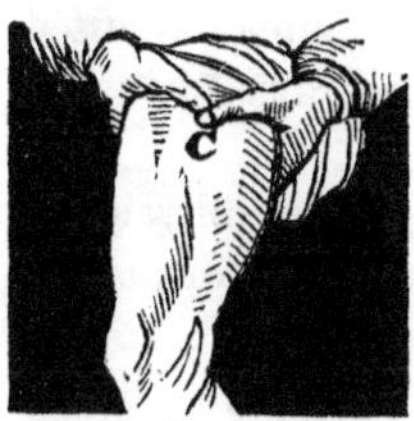

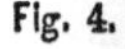
Fig. 4.

Fig. 5.

The artery in the arm runs down the inner side of the large muscle in front, quite close to the bone; lower down it gets farther forward toward the bend of the elbow. It is most easily found and compressed a little above the middle, (*see Figure* 5.)

Care should be taken to examine the limb from time to time, and to lessen the compression if it becomes very cold or purple; tighten up the handkerchief again if the bleeding begins afresh.

A BRIEF HISTORY OF THE STEAM-ENGINE.

Hero of Alexandria, who flourished about 200 B. C., has left us a description of a steam-engine by which machinery could be set in motion.

In the year 450 A. D., Anthemius, an architect,

tried some experiments with steam, which is the second attempt on record to use steam as a motive-power.

In the year 1543, Blasgo de Garay, a Spaniard, propelled a vessel of 200 tons, in the harbor of Barcelona, by the force of steam.

In the year 1615, De Caius, a Frenchman, devised a machine by which water could be raised in tubes through the agency of steam.

In the year 1630, Branca, an Italian physician, ground his drugs by means of a wheel set in motion by steam.

In 1656, the Marquis of Worcester made some very important improvements in the steam-engine.

In the year 1685, some experiments were made with steam by Pepin, a Frenchman, who devised the mode of giving a piston an up and down movement by alternately generating and condensing the steam in a cylinder.

In the year 1698, Savery, an Englishman, constructed a steam-engine superior to any before invented.

In the year 1705, Newcomen made some very important improvements in the steam-engine. To Newcomen belongs the honor of building the first engine that bore any resemblance to modern steam-engines.

In the year 1764, James Watt constructed the first perfect stationary steam-engine that was ever made up to that date.

In the year 1769, Nicholas Cugnot constructed the first self-moving locomotive-engine.

In the year 1786, the Marquis Jouffry constructed a small steamboat on the Saone.

In the year 1789, William Symington made a voyage in a small steamboat on the Firth of Clyde.

In the year 1782, Ramsey propelled a boat by steam in New York harbor.

In the year 1788, John Fitch, of Philadelphia, navigated a boat by steam on the Delaware River.

In the year 1793, Robert Fulton made some important experiments with steam.

In the year 1794, Oliver Evans, a native of Philadelphia, constructed the first locomotive or road-steamer in America.

In the year 1803, John C. Stevens, of New York, obtained a patent for a steamboat.

HISTORY OF THE DIFFERENT PARTS OF THE STEAM-ENGINE IN DETAIL.

The inventor of the slide-valve is unknown: it was spoken of by Hero 2,000 years ago.

The valve was made self-acting by a boy named Humphrey Potter.

The eccentric was invented by Murdock, as was also the crank.

The beam and connecting-rod were invented by Newcomen.

The radius and parallel bars were invented by Watt.

BOILERS.

John C. Stevens invented the tubular boiler, 1804.
James Neville invented the multi-tubular, 1826.

VOCABULARY.

Anchor Brace. — A brace for supporting brick walls around boilers.

Ash Pit. — The opening under the grate-bars, through which air is admitted to the furnace.

Back Rest. — A rest used for the piston, in cases where the piston-rod passes through both cylinder-heads.

Baffle. — A screen in steam-boilers, to prevent water from passing over into the cylinder.

Balance Wheels. — A wheel on the main shaft of the engine to insure uniform motion.

Bed Plate. — The iron plate or base of the engine, resting on the foundation.

Blow-off Cock. — A cock at the bottom of the boiler, through which to empty the boiler.

Boiler. — The source of power; the vessel in which the steam is generated.

Bonnet. — The cover of the steam-chest.

Bridge Wall. — The back wall of the furnace.

Buck Stave. — A brace to sustain the brick work around a boiler.

Cam. — See ECCENTRIC.

Check Valve. — The valve between the pump and boiler, to prevent the back pressure of the water.

Chimney. — The flue through which the smoke escapes from the furnace.

Connecting Rod. — A rod to communicate the pressure on the piston to the crank.

Crank. — The part of the engine that converts rectilinear motion of the piston into rotary motion.

Crank Pin. — The bearing by which the connecting-rod is attached to the crank.

Cross Head. — A block moving in guides, having the end of piston secured within it at one end and attached to the connecting-rod at the other.

Crow Feet. — Braces used in steam-boilers.

Crown Bars. — Bars attached to the upper side of the crown-sheet to prevent springing.

Crown Sheet. — The top sheet of the furnace directly over the fire.

Cut Off. — An additional valve to cut off steam in the cylinder at any point of the stroke.

Cylinder. — The vessel or tube in which the piston-head moves.

Cylinder Heads. — The back and front ends of the cylinder.

Damper. — A door to exclude the air from the furnace.

Dome. — An elevated chamber on top of the boiler, from which the steam is taken.

Drip Cocks. — Small cocks in the cylinder-heads, through which the condensed water escapes.

Driving Belts. — The belt that transmits the power of the engine to the machinery.

Driving Pulley. — A pulley on the main shaft of the engine.

Eccentric. — The cam that gives movement to the valves.

Eccentric Straps. — The straps that surround the eccentric, and to which the rod is attached.

Eduction Port. — A passage on the side of the cylinder to lead away the exhaust steam.

Exhaust Pipe. — The pipe through which the steam escapes from the cylinder.

Expansion Valve. — A cut-off.

Feed Pipe. — A pipe through which feed-water is fed into the boiler.

Fire Box. — The furnace of a locomotive boiler.

Flues. — Passages through which the heat is conducted in flue boilers.

Foaming. — An artificial excitement or boiling, caused by the water becoming greasy or foul.

Follower Bolts. — Bolts that secure the follower-plate to the piston-head.

Follower Plate. — The plate that covers the packing on the piston-head.

Foundation. — The brick or stone base on which to set the engine.

Foundation Bolts. — The bolts by which the bed-plate is made fast to the foundation.

Furnace. — A chamber in which the coal is burned under a boiler.

Gasket. — A packing for the man-hole of the boiler.

Gauge Cocks. — Cocks at different levels, to ascertain the height of the water in the boiler.

Gibb. — The counter part of a key, used with the key to take up lost motion in boxes.

Gland. — A bushing to secure the packing in the stuffing-boxes.

Glass Gauge. — A glass tube on the front of a boiler, to indicate the height of the water.

Governor. — A machine to regulate the speed of an engine.

Grate. — Parallel bars that support the fuel in the furnace.

Grommet. — A ring of hemp, a packing.

Guides. — The bars on which the cross-head moves.

Gusset Stay. — A stay used in boilers.

Hand Hole. — See MUD HOLE.

Heater. — A cylinder in which to heat the feed-water before entering the water.

Housing. — The frame of vertical engines.

Induction Ports. — The passages through which the steam is admitted to the cylinder.

Injector. — An instrument used for supplying feed-water to boilers.

Jacket. — A covering for steam-cylinders.

Journals. — The part of the engine-shaft resting in the pillow-block boxes.

Key. — A wedge to take up lost motion in boxes on crank-pin and cross-heads.

Lagging. — A wooden sheathing around the cylinder or steam-pipe.

Lap. — The distance which the valve overlaps the steam-ports.

Lead. — The distance which the induction port is open when the piston commences the stroke.

Link Motion. — An arrangement for reversing the motion of engines.

Lubricator. — The cup or valve through which the oil or tallow is admitted to the cylinder.

Man Hole. — A hole to admit a man within the boiler.

Man-Hole Brace. — A cross brace to secure the plate in the man-hole.

Man-Hole Plate. — A plate to close the man-hole.

Midfeather. — A brace under a boiler.

Mud Hole. — A small opening at the bottom of the boiler, through which to remove the mud.

Packing. — A substance used to make a steam-joint around the piston-rod.

Packing Hook. — A steel hook, used for removing packing from stuffing-boxes.

Packing Rings. — The rings that form a steam-tight joint in the cylinder.

Packing Stick. — A small stick used to drive packing into the box.

Pitman. — A connecting-rod.

Pet Cock. — A small cock, between the check valve and pump, used to see if the pump is working.

Pipe (in Mechanics). — A metallic tube.

Piston Head. — The head on which the rings fit in the cylinder.

Piston Rod. — A rod secured at one end to the piston-head and at the other to the cross-head.

Plunger. — A solid piston of a pump.

Ports. — Openings or passages for the admission of steam or water.

Priming. — The passage of water along with the steam into the cylinder.

Pump. — A machine used for lifting or forcing water.

Rocker. — An intermediate bearing between the eccentric and valve rods.

Rocker Shaft.—The shaft on which the rocker vibrates.

Safety Valve. — A valve on the boiler, to discharge surplus steam.

Shackle Bar. — A connecting-rod.

Shell. — Outside sheets of a boiler.

Slice Bar. — A bar used for the purpose of cleaning fires in boiler furnaces.

Slide Valve. — The valve that admits the steam to the cylinder of slide-valve engines.

Stack. — See CHIMNEY.

Spider. — A piston-head.

Stagger Plate. — A plate to prevent the passage of water with the steam from the boiler to the cylinders.

Starting Bar. — A bar used to move the valve of an engine when the valve-gear is disconnected.

Steam Chest. — A box on the top or side of the cylinder containing the valve which admits steam to the piston.

Steam Gauge. — A gauge to show the pressure of steam on a boiler.

Steam Pipe. — The pipe through which the steam flows from the boiler to the steam-chest.

Stop Valve. — The valve that closes the communication between the boiler and steam-chest.

Straps. — The pieces that secure the crank-pin and cross-head boxes to the connecting-rod.

Stroke. — The distance travelled by the piston at each revolution of the crank.

Stub Ends. — The ends of the connecting-rod that butt against the boxes.

Stuffing Box. — A chamber to receive the packing, where rods or spindles having an end motion require to be made steam-tight.

Thimble. — An iron ring or ferrule used for stopping leaks in tubes of steam-boilers.

Tubes, — of a boiler through which the heat and smoke escape to the chimney.

Tube Sheets. — The ends of the boiler or sheets to which the tubes are attached.

Union. — A steam-tight joint formed with a coupling.

Water Gauge. — See GLASS GAUGE.

Wrist. — The bearing in the cross-head on which the connecting-rod vibrates.

THE END.

www.ingramcontent.com/pod-product-compliance
Lightning Source LLC
LaVergne TN
LVHW011205110826
845150LV00006B/1331

* 9 7 8 1 4 2 5 5 1 8 5 2 3 *